Natural Environment Research Council
INSTITUTE OF TERRESTRIAL ECOLOGY

Distribution and status
of bats in Europe

D1339456

R E Stebbings
and Francesca Griffith

Monks Wood Experimental Station
Abbots Ripton
Huntingdon
Cambs PE17 2LS

Printed in Great Britain by
Cambrian News (Aberystwyth) Ltd

Published in 1986 by
Institute of Terrestrial Ecology
Administrative Headquarters
Monks Wood Experimental Station
Abbots Ripton
Huntingdon
PE17 2LS

BRITISH LIBRARY CATALOGUING-IN-PUBLICATION DATA
Stebbings, R. E.
    Distribution and status of bats in Europe
    1. Bats
    I. Title II. Griffith, F. III. Institute of Terrestrial Ecology
    599.4 QL737. C5

ISBN 0 904282 94 5

ACKNOWLEDGEMENT
This work was funded in part by the Nature Conservancy Council
and in part through the NCC by an EEC contract (No. U 81 524).

COVER ILLUSTRATION
Eptesicus serotinus — serotine bat
(All photographs by R E Stebbings)

The *Institute of Terrestrial Ecology (ITE)* was established in 1973,
from the former Nature Conservancy's research stations and staff,
joined later by the Institute of Tree Biology and the Culture Centre of
Algae and Protozoa. ITE contributes to, and draws upon, the
collective knowledge of the 14 sister institutes which make up the
Natural Environment Research Council, spanning all the environmental sciences.

The Institute studies the factors determining the structure, composi-
tion and processes of land and freshwater systems, and of individual
plant and animal species. It is developing a sounder scientific basis
for predicting and modelling environmental trends arising from
natural or man-made change. The results of this research are
available to those responsible for the protection, management and
wise use of our natural resources.

One quarter of ITE's work is research commissioned by customers,
such as the Department of Environment, the Commission of the
European Communities, the Nature Conservancy Council and the
Overseas Development Administration. The remainder is fun-
damental research supported by NERC.

ITE's expertise is widely used by international organizations in
overseas projects and programmes of research.

Dr R E Stebbings
Institute of Terrestrial Ecology
Monks Wood Experimental Station
Abbots Ripton
Huntingdon
PE17 2LS
048 73 (Abbots Ripton) 381

CONTENTS

INTRODUCTION

National and international agencies, societies and individuals are increasingly seeking information on the current distribution and status of bats in Europe. The stimulus to produce this report came in 1980 when the Environment and Consumer Protection Service of the Commission of the European Communities (EEC) requested information on the threatened flora and vertebrate fauna within the EEC. This was needed so that there could be a co-ordinated approach to the conservation of these biota. The EEC had become aware of its commitment to international conservation in 1979 when the Bern Convention on the Conservation of European Wildlife and Natural Habitats was published and presented by the Council of Europe to its 21 Member States for signing and ratification.

Despite one third of the indigenous terrestrial mammal species in Europe being bats, little is known of their detailed distribution and numbers. However, accumulation of knowledge has gathered momentum and, although much has been published, it is widely scattered, often in obscure journals. Brink in the early 1950s (1967) was one of the first to attempt a compilation of distribution maps for the European bats and these have been reproduced many times. However, present knowledge is much improved and new maps are needed.

This report is an attempt to summarize the present status of bats in 27 countries in western Europe and provides sources where more detailed information can be found. I hope it will also stimulate more systematic recording and detailed ecological research, which are necessary to establish conservation requirements for each species.

Many bats move between summer and winter roosts and these may be separated by large distances. A few species have significant populations in EEC countries in winter (eg *Vespertilio murinus*) but in summer these bats form nursery colonies in north-east Europe, particularly in Poland and the USSR. Thus, it is vital to consider the range and migrations of a species when preparing conservation strategies (Strelkov 1969; Roer 1971).

METHODS

A draft was prepared by consulting about 450 papers giving information on distribution, ecology or status. Over half are quoted in the bibliography and readers should additionally consult the 2 European bat journals, *Myotis* and *Nyctalus*, which contain many more papers and references. An arbitrary assessment was made as to whether the details were valid and, if so, how the information should be presented. Decisions were derived from the sum total of knowledge at the time, and hence re-assessment was constantly necessary.

The draft was presented at the first European Symposium on Bat Research, held at Bonn, 16–20 March 1981. Some of the 120 scientists from 16 European countries provided comments at the meeting and 38 copies of the draft were sent to scientists in 17 countries for more detailed comments and additions. Thirty-nine scientists from 17 States (see Acknowledgements) provided corrections, some collectively, and some providing details for 2 or more countries. Although most information was complementary, there were contradictory opinions, particularly regarding status and distribution of individual species. We have used our judgement in presenting the information here and we are entirely responsible for any errors.

Opinions about status are inevitably subjective, especially for those species which are difficult to find, such as tree-roosting bats. Within Europe there are only a few small areas where detailed work allows quantitative statements. However, in many highly agricultural and urbanized areas, observers often have the impression that declines are far larger than is apparent from recent studies. Historic records are usually lacking but sometimes former colony size can be estimated, for example by measuring the area covered by guano piles and relating that area to measured densities of bats in clusters.

DISTRIBUTIONAL AREAS
Distributions have been plotted in all European countries east to the USSR. However, detailed assessment of the status of bats was confined to western Europe to the eastern borders of Finland, Poland, Czechoslovakia, Hungary, Yugoslavia and Greece. Partial information is given for Rumania and Bulgaria. No species is totally limited to within western Europe, but for many a large proportion of their range is included.

INTERPRETATION OF MAPS
The maps show the areas where bats may be caught either roosting, feeding or migrating. Species which are widely distributed may differ greatly in abundance; eg, *Pipistrellus pipistrellus* (map 21) is a common bat over most of Europe with numbers probably totalling several millions, but *Pipistrellus nathusii* (map 22), although found widely in countries bordering the Atlantic and North Sea, may total only a few hundred bats.

SPECIES ACCOUNTS
Since 1960, 3 new bat species have been identified and described, and

6

others are awaiting description. These species are morphologically very similar to other well-known forms. In other areas of the Palaearctic where many of our species occur, even less is known of their systematics and no doubt other species remain to be identified.

NOMENCLATURE

The classification and scientific nomenclature follows Corbet (1978) and selected, widely used vernacular names are given when available in English (E), French (F), German (G), Danish (D), Dutch (N), Italian (I), Spanish (S) and Greek (H). Additional local names were published by Roer and Hanak (1971).

HABITAT

Little is known of the habitat requirements of most species. Although formerly living in natural habitats, many successfully adapted to and probably became more abundant in man-made habitats and particularly adopted buildings, tunnels and mines for roosting. Changes in climate may also affect distribution and abundance, but we do not have any reasonable idea of what changes happened historically, or are occurring today. Generally, bats are found in all types of habitat, excepting the extreme arctic regions and alpine peaks. Greatest abundance and species diversity are found in sheltered, mixed habitats, including woodland, pasture and riparian habitats.

POPULATIONS

About half of Europe's species have not been studied, even in a single site, and knowledge about the others is very limited and patchy. A few of the easily found species were studied from the 1930s and these projects often led to the disappearance of those bats. The sensitivity of bats to disturbance has only been well known since the early 1960s. Despite intensive study of some species in local areas, often showing immense declines, we have no precise information on overall population trends. However, from our sparse ecological knowledge of these bats, we can assume similar declines would have occurred elsewhere. With the present level of bat observation in most countries, even very large changes in population sizes will remain undetected. Usually the cause of a population change is unknown, but one colony of 7000 *Miniopterus schreibersii* present in 1950 in the west of France virtually disappeared in 10 years due to human disturbance (Brosset 1966). A large survey in Britain of house bats showed a decline of over 50% between 1978 and 1980 (Stebbings & Jefferies 1982; Stebbings & Arnold 1982). It was thought that adverse weather in late spring was the major cause. Population recovery can only be slow because most bats produce a maximum of one young per year and some, like *Rhinolophus ferrumequinum*, are at least 4 years old before breeding. Other studies have shown that some bats do not breed every year.

THREATS

Some bats are highly colonial and, although a few colonies may be known containing thousands of bats, they may represent the entire breeding stock covering thousands of km^2. For example, a *Rhinolophus ferrumequinum* colony in Wales totalling about 400 bats occupies an annual area of 2500 km^2 (Stebbings' data). Thus, catastrophes to breeding colonies where almost all adult females congregate can remove bats from wide areas instantly. This applies equally to colonies in natural sites, eg caves and hollow trees, and man-made sites, eg buildings and mines. Catastrophes can be accidental, eg flooding or collapse of caves and mines, trees blowing down, or burning down of buildings, killing by vandals and collectors; but almost certainly of greater significance today is remedial timber treatment in buildings. This latter practice has been increasing since 1950 and is probably the most important factor killing bats and reducing breeding success. In Britain, the estimated number of dwellings treated annually rose from

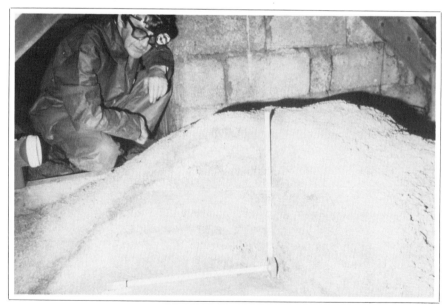

Plate 1. Greater horseshoe bat nursery colony. A very large (1500 bats) colony was killed with the insecticide dieldrin. The pile of droppings was 49 cm deep

35 000 in 1972 to 100 000 in 1979 and 500 000 in 1983 (Stebbings' data). Many chemicals used in the treatment (mostly gamma HCH (Lindane) and dieldrin) are extremely lethal to bats (Plate 1). They remain on the surface of treated timber for many years and volatilize slowly, and, because bats roost on these surfaces, the chemicals are absorbed through the skin, mouth and lungs. Bats are now so dependent on buildings that their future existence may be under threat, unless different methods of timber preservation are adopted. Bats in the Netherlands have probably suffered most in this way (Daan *et al.* 1980; Voûte 1981) but there is little documentation elsewhere. Many bats roost in hollow walls of buildings, and cavity wall insulation must be both killing bats and preventing access.

Other major threats to bats involve their food and feeding habitat, and generally we know little of their requirements. European bats are insectivorous, and are found in almost all habitats. Some catch insects emerging from water, others feed largely on the ground or in free air, well away from trees, and still others glean food off bark and foliage. The reduction in insect abundance due to water pollution, changing agricultural practice and insecticides must be having significant effects on survival and breeding

Plate 2. St Pietersberg stone mines on the Netherlands/Belgium border are being destroyed by open cast mining of limestone for cement manufacture. Tunnels contain one of the most important hibernating populations of bats in Europe.

success in bats, but again we have no precise measure of such effects. The recent change from hay making to silage prevents the maturation of some insects and is probably reducing insect abundance and variety. From experimental work, we know that bats may be more sensitive to organochlorine pesticides than some other higher vertebrates (Jefferies 1972; Clark 1981). More insidious changes include loss of roost sites, for example by the tidying up of ancient monuments where small gaps are infilled and cellars or tunnels have their entrances blocked by doors rather than grilles. Commercialization of caves often leads to depopulation of bats and more mines are now being developed as shelters from nuclear attack. Mines which have often been used by bats for hundreds of years are being blocked or capped, or destroyed by open cast quarrying (Plate 2).

CONSERVATION MEASURES

Legal protection of bats is afforded to all species in all EEC countries. Bats are also protected in all other European countries, but the level of protection varies greatly between States. Also variable is the degree of implementation and public awareness in each country. Legal protection is most useful for its educative value, providing it is publicised, but it does not act as a deterrent to prevent vandalistic killing, or casual disturbance by tourists or speleologists. Effective conservation will best be assured by careful management of forestry, agricultural and riparian habitats with minimal use of pesticides. A most urgent need is to find new methods of dealing with insect and fungal infestation of buildings, ie remedial timber treatment. All new timber used in buildings should be pre-treated, which in the long term is much cheaper and more effective than remedial treatment. Also, new chemical preparations should be formulated which will be less damaging to bats. Such formulations will involve some research but they could probably be produced with little difficulty and probably little more expense than existing products. Legislation will be required in all countries making these formulations mandatory in remedial timber treatment. Adoption of this measure should be a priority task in each country and is likely to have greatest benefit to the long-term conservation of bats. Research has already shown that the insecticide Permethrin is effective at controlling wood-boring beetles and does not appear to harm bats (Berry 1983; Racey & Swift 1986).

In many States, individual bat roosts have been specifically protected, eg caves, mines and ancient monuments, and this sort of protection is now being afforded to some dwelling houses, hollow trees and, exceptionally, to general habitat on which significant bat populations depend. Many countries have schemes to provide bat roost boxes (artificial trees holes) and these are very successful in attracting and possibly increasing bats in some habitats. Blocked caves and mines are being re-opened for bats and digging new tunnels is being suggested. Disused railway and other tunnels can be made into suitable bat hibernacula by management of the internal climate.

RECOMMENDATIONS

An education programme is required in each country to explain the value of bats and to dispel the fear, superstitions and hostility that is often directed to bats by the general public. The programme needs to create awareness in all levels of society of bats' requirements, and this will help to ensure their long-term conservation.

All countries should implement measures to protect bats.

Adequate protection should be given to all breeding colonies, especially the prevention of disturbance of cave colonies by speleologists and tourists. Priority should be given to endangered and rare species.

More research is required, particularly for the endangered species, to find what are the critical habitats and causes of declines.

Research should be done to develop treatments for insect and fungal infestations of buildings which are not harmful to roosting bats, and these treatments should become mandatory for all remedial timber work in buildings.

ACKNOWLEDGEMENTS

Many people kindly provided help, advice and information during the preparation of this report. Rosemary Parslow kindly collated information on some of the endangered species. The following provided detailed comments and information, whose help is gratefully acknowledged.

EEC countries

Belgium
Dr J Fairon
Dr R Jooris

Denmark
Dr Hans J Baagøe
Dr Birger Jensen

Federal Republic of Germany
Dr Friedel Knolle
Dr Anton Kolb
Dr Alfred Nagel
Dr Hubert Roer
Karl-Hans Taake

France
J F Noblet
Jean-Louis Rolandez
Dr Yves Tupinier

Italy
Dr Edoardo Vernier

Netherlands
Dr Gerhard H Glas
E De Grood
J M van den Hoorn
P H C Lina
Dr Aldo M Voûte

Other European countries

Bulgaria
Dr P Beron

Czechoslovakia
Prof Jiři Gaisler
Dr Vladimir Hanák
Dr Ivan Horáček
Dr Petr Rybář

Democratic Republic of Germany
Dr Joachim Haensel

Hungary
Dr György Topál

Norway
Dr Jørgen A Pedersen

Poland
Wieslaw Bogdanowicz
Dr A Krzanowski
Dr Andrzej L Ruprecht
Dr B W Woloszyn

Rumania
Dr P Barbu

Sweden
Dr Ingemar Ahlén
Dr Rune Gerell
Dr Olof Ryberg

Republic of Ireland
Patrick O'Sullivan
Patrick Warner

Switzerland
Prof Villy Aellen
Dr Peter E Zingg

Yugoslavia
Prof Beatrica Dulić

Comments in reports or letters from correspondents are referred to by name, followed by month/year.

REFERENCES

Berry, R. W. 1983. Recent developments in the remedial timber treatment of wood-boring insect infestations. In: *Biodeterioration 5*, edited by T. A. Oxley & S. Barry, 154-165. Chichester: Wiley.

Brink, F. H. van den. 1967. *A field guide to the mammals of Britain and Europe.* London: Collins. (First published in 1955 in the Netherlands).

Brosset, A. 1966. *La biologie des Chiroptères.* Paris: Masson.

Clark, D. R. 1981. Bats and environmental contaminants: a review. *Spec. scient. Rep. U.S. Fish Wildl. Serv., Wildl.,* no. 235.

Corbet, G. B. 1978. *The mammals of the Palaearctic region: a taxonomic review.* London: British Museum; Ithaca: Cornell University Press.

Daan, S., Glas, G. H. & Voûte, A. M. eds. 1980. De Nederlandse vleermuizen: Bestandsontwikkelingen in winter-en zomerkwartieren. *Lutra,* **22,** 1-118.

Jefferies, D. J. 1972. Organochlorine insecticide residues in British bats and their significance. *J. Zool.,* **166,** 245-263.

Racey, P. A. & Swift, S. 1986. The residual effects of remedial timber treatments on bats. *Biol. Conserv.,* **35,** 205-214.

Roer, H. 1971. Weitere Ergebnisse und Aufgaben der Fledermausberingung in Europa. *Decheniana - Beihefte,* **18,** 121-144.

Roer, H. & Hanák, V. 1971. Glossarium vespertilionum Europae. *Myotis,* **8,** 9-27.

Stebbings, R. E. & Arnold, H. R. 1982. Bats - an insecticide under threat? *Nat. Devon,* **3,** 7-26.

Stebbings, R. E. & Jefferies, D. J. 1982. *Focus on bats: their conservation and the law.* London: Nature Conservancy Council.

Strelkov, P. P. 1969. Migratory and stationary bats (Chiroptera) of the European part of the Soviet Union. *Acta zool. cracov.,* **14,** 393-439.

Voûte, A. M. 1981. The conflict between bats and wood preservatives. *Myotis,* **18-19,** 41-44.

SUMMARY OF THE BATS OF EUROPE

Species			Status	
			EEC	World
	Family Rhinolophidae			
1.	*Rhinolophus ferrumequinum*	- Schreber	E	E
2.	*Rhinolophus hipposideros*	- Bechstein	E	E
3.	*Rhinolophus blasii*	- Peters	E	K
4.	*Rhinolophus euryale*	- Blasius	E	V
5.	*Rhinolophus mehelyi*	- Matschie	E	R(?E)
	Family Vespertilionidae			
6.	*Myotis bechsteinii*	- Kuhl	R(?E)	R(?E)
7.	*Myotis nattereri*	- Kuhl	V	V
8.	*Myotis capaccinii*	- Bonaparte	V(?E)	K(?E)
9.	*Myotis dasycneme*	- Boie	E	E
10.	*Myotis daubentonii*	- Kuhl	?NT	?NT
11.	*Myotis nathalinae*	- Tupinier	K	K
12.	*Myotis emarginatus*	- Geoffroy	E	E
13.	*Myotis mystacinus*	- Kuhl	V	?NT
14.	*Myotis brandtii*	- Eversmann	V	?NT
15.	*Myotis blythi*	- Tomes	E	E
16.	*Myotis myotis*	- Borkhausen	E	E
17.	*Barbastella barbastellus*	- Schreber	V(?E)	V
18.	*Plecotus auritus*	- Linn	V	V
19.	*Plecotus austriacus*	- Fischer	V	V
20.	*Miniopterus schreibersii*	- Kuhl	E	?E
21.	*Pipistrellus pipistrellus*	- Schreber	V	V
22.	*Pipistrellus nathusii*	- Keyserling & Blasius	R	V
23.	*Pipistrellus kuhli*	- Kuhl	V	V
24.	*Pipistrellus savii*	- Bonaparte	V	V
25.	*Eptesicus serotinus*	- Schreber	NT	NT
26.	*Eptesicus nilssonii*	- Keyserling & Blasius	R	K or NT
27.	*Vespertilio murinus*	- Linn	R	R
28.	*Nyctalus leisleri*	- Kuhl	R(?V)	V
29.	*Nyctalus noctula*	- Schreber	V(?E in NW)	V
30.	*Nyctalus lasiopterus*	- Schreber	R	R
	Family Molossidae			
31.	*Tadarida teniotis*	- Rafinesque	V	V

E = Endangered   K = Insufficiently known   NT = Not threatened
R = Rare   V = Vulnerable

Tables 1a and b show the occurrence and overall status of each species in the European countries covered by this report.

A blank space indicates the species does not occur, or at least it is unlikely and there is no positive information. It should be remembered that some species fly long distances between roosts (hundreds of kilometres) and all species will cross State boundaries.

Table 1a Occurrence and status of bats in EEC countries

	Republic of Ireland	Denmark	Netherlands	Belgium	Luxembourg	France	Federal Rep. of Germany	Italy	Greece	Spain	Portugal
Rhinolophus ferrumequinum		E	X	E	E	E	E	E	E	E	E
R. hipposideros	V	E		E	E	E	E	E	E	E	E
R. euryale						E		E	E	E	E
R. mehelyi						E		E	E	E	E
R. blasii								E	E		
Myotis mystacinus	V	V	V	V	E	V	V	V	V	V	V
M. brandtii	?	V	V	V	E	V	V	?	?	?	?
M. emarginatus			V	E	E	E	E	E	V	E	
M. nattereri	V	V	V	K	R	V	V	V	?	V	V
M. bechsteinii		R		R	R	R	N	R		R	R
M. myotis (M)		E		E	E	E	E	E	E	E	E
M. blythi (M)						E	X	E	E	E	E
M. daubentonii	V	V	N	N	V	V	V	V		V	V
M. nathalinae						V		?		V	?
M. capaccinii						E		V	?	V	
M. dasycneme			V	E	E	E	E	R			
Pipistrellus pipistrellus	V	V	N	N	N	V	N	V	V	V	V
P. nathusii (M)		R	K	R	R	R	R	R	R	R	R
P. kuhli						V		V	V	V	V
P. savii						V	V	V	V	V	V
Nyctalus leisleri (M)	V	V		K	R	?	V	V	V	V	V
N. noctula (M)		E	V	K	V	V	V	V	V	V	V
N. lasiopterus (M)						R		R	R	R	
Eptesicus nilssonii		K				R	R	R			
E. serotinus		V	N	N	V	V	V	V	V	V	V
Vespertilio murinus (M)		N				R	R	R	R		
Barbastella barbastellus		R	R	R	E	E	V	E	V	R	R
Plecotus auritus	V	V	V	V	V	V	V	V	?	V	
P. austriacus		R		K	V	V	V	V	V	V	V
Miniopterus schreibersii (M)						E		E	E	E	E
Tadarida teniotis						R		V	V	V	V

X = Extinct   V = Vulnerable   (M) = Migratory
E = Endangered   N = Not threatened   R = Rare
K = Occurs but there is no exact information as to status
? = Where bats probably occur–but there are no records

Table 1b Occurrence and status of bats in non-EEC countries

	Norway	Sweden	Austria	Switzerland	Liechtenstein	Malta	Democratic Rep. Germany	Poland	Finland	Czechoslovakia	Hungary	Yugoslavia	Albania	Bulgaria	Rumania
Rhinolophus ferrumequinum			E	E	E	E	E	X		E	V	E	E		K
R. hipposideros			E	E	E	?	E	E		E	E	E	E		K
R. euryale			E	X	E	?				E	E	E	E	K	
R. mehelyi						?				E	E	K	K		
R. blasii						?				E	E	R			
Myotis mystacinus	N	N	V	V	V	V	V	V	V	V	E	V	V		K
M. brandtii	V	K	?	V	?	?	V	V	V	V	E	?	?		
M. emarginatus			E	E	E	?	E	X		E	E	E	E		K
M. nattereri	R	V	V	V	V	?	V	V	R	V	E	R	?	K	K
M. bechsteinii		E	R	R	R	?	R	R		R	E	R		R	K
M. myotis (M)			E	E	E	E	E	E		E	V	E	E		K
M. blythi (M)			E	E	E	E				E	V	E	E		K
M. daubentonii	N	N	V	V	V		V	V	N	N	V	V	V		K
M. nathalinae			?	V	?		?	R							
M. capaccinii			X	X	X					X	E	E	K		
M. dasycneme		K					E	R		R	E				K
Pipistrellus pipistrellus	N	N	V	V	V	V	V	V		V	N	N	N		K
P. nathusii (M)		K	R	V	R		V	V		R	E	R	R		K
P. kuhli			V	V	V	V						N	N	K	K
P. savii			V	V	V	V						N	N	K	K
Nyctalus leisleri (M)			V	V	V	V	V	V		R	E	R	R	K	K
N. noctula (M)	K	V	V	V	V	?	V	V	R	V	N	V	V		
N. lasiopterus (M)			R	R				R		R	E	R	?	R	
Eptesicus nilssonii	N	N	R	V	R		R	R	N	V	E	R			
E. serotinus	K	R	V	V	V		V	V		V	N	N			K
Vespertilio murinus (M)	N	N	R	R	R		V	R	R	R	E	R			K
Barbastella barbastellus	R	V	V	V	V		R	V		V	E	E		K	
Plecotus auritus	N	N	V	V	V		V	V	N	V	E	R	?		K
P. austriacus		K	V	V	V	V	V	R		V	V	V	V		
Miniopterus schreibersii (M)			E	E	E	?				E	V	N	N		K
Tadarida teniotis			R	R	R	V						R	?		

SPECIES ACCOUNTS

		EEC	World
CHIROPTERA: RHINOLOPHIDAE	Status:	E	E

1. *Rhinolophus ferrumequinum* - Schreber 1774

E. Greater horseshoe bat
F. Grand rhinolophe fer à cheval
G. Gros Hufeisennase
I. Ferro di Cavallo maggiore

D. Stor hesteskonaese
N. Grote hoetijzerneus
S. Murciélago grande de herradura
H. Rhinolophos i megali

DISTRIBUTION

EEC: SW Britain, SE Belgium, France, Luxembourg, S Germany, Italy, Sicily, Corsica, Sardinia, Greece. Now probably extinct in Netherlands where it was formerly found in the South Limburg Province (Glas, personal communication).

Europe: Found throughout Hungary, Bulgaria, Austria, Switzerland, Yugoslavia, Albania, Rumania, Spain and Portugal.

Czechoslovakia: Northern boundary of species distribution, found throughout the year in south Slovakia, especially in the Slovak Karst, and occasionally in north Slovakia, Moravia and Bohemia (182).

Poland: Only one known record from the Nietoperzowa cave near Ojcow in the Krakow-Czestochowa Uppland (1962) (Bogdanowicz, pers. comm., Bogdanowicz & Ruprecht, pers. comm., 114).

Rumania: Found in limestone caves in the Rumanian Dobrogea adjacent to the Black Sea (43).

World: Occurs in the entire southern Palaearctic from Britain to Japan with the northern boundary through S England, central Germany, Crimea, Caucasus, Kopet Dag, Kirgizia, S Korea, Hokkaido and Honshu. The southern boundary includes Morocco, Algeria and Tunisia, Palestine, Iran, Afghanistan and through Himalayas to Yunnan.

Rhinolophus ferrumequinum

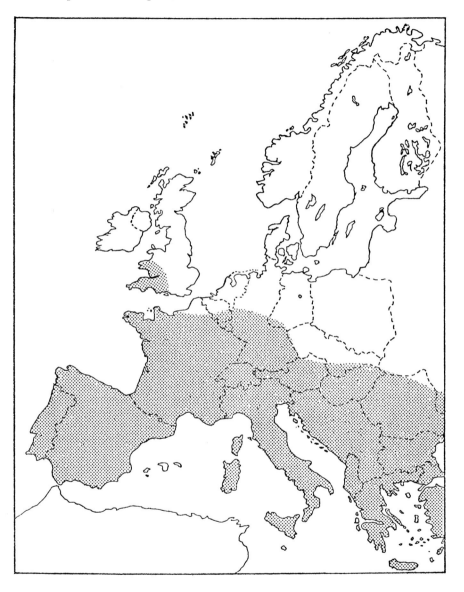

HABITAT

Associated mostly with woods, scrubland and grasslands, often near water. Feeds on a wide variety of mostly large insects in woodland and low over old pasture. Nursery roosts in summer are mostly in buildings, but also in caves and mines, particularly in southern areas, and it mostly hibernates in caves and similarly humid places in winter, eg cellars, icehouses and mines.

Exceptionally, some have been captured in cracks in trees during spring (219).

POPULATION

EEC: Very large declines in numbers have been recorded (eg Britain 98%) in the past century and the species range is also reduced. It is declining rapidly in southern Belgium and Luxembourg, as well as other areas in the north. Large colonies are still found in southern and SE Europe, although they too are declining.

Britain: It has become extinct in several counties in central southern and SE England, and in 1983 the British population numbered about 2200 (a decline of over 98% in a century) (202).

France: Widespread in Isère department (157) and Ain region (southern zone of the Jura Mountains) (Rolandez, pers. comm. 5/1981), found throughout the Rhône-Alpes region (219). Generally, bats have been seen individually or in small groups of about 10, but have been observed during the winter in dense colonies of more than 100 individuals (disused mines in Beaujolais) (219). In the Isère department, 1415 were ringed between 1936 and 1960 (157).

Netherlands: A total of 516 individuals were ringed between 1936 and 1951 in the South Limburg mines. A breeding cluster was found in St Pietersberg mines in 1939 numbering 75 bats, and 80 in 1940. The species is now extinct (23).

Europe: **Czechoslovakia:** In summer, nursery colonies of 50–100 individuals are found; in winter, usually only isolated individuals, but occasionally up to 20–100 bats in groups. A slight decrease in population numbers is occurring (182).

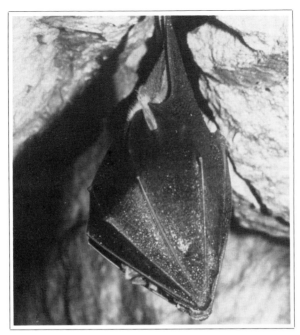

Rhinolophus ferrumequinum - greater horseshoe bat

Rumania: Small numbers have been recorded in the Rumanian Dobrogea from 1956 onwards. In 1979 a breeding colony was found numbering 200–300 bats. Both summer and winter colonies occur (43).

Spain: Breeding colonies have been recorded at Arredondo, La Saja and San Leonardo. Colonies were also found in the Cantabrique Mountains and Almadenejos. A colony occurs during the summer in the cave at Cubera, numbering up to 50 bats of both sexes. In the Ciudad-Real Province, 2 breeding colonies are known numbering over 100 bats (221).

World: In Israel the species is virtually extinct (Makin, pers. comm. 1979) and it appears to be declining rapidly elsewhere. The systematics of the species is not clear and, although it is apparently widespread, it may not be the same species throughout the range described above.

THREATS

Disturbance in caves, collection for research, remedial timber treatment in buildings, exclusion from buildings, vandalistic killing, habitat change involving the loss of large insects, particularly the beetles *Melolontha* and *Geotrupes*, are the major threats and causes of declines.

Also, because the bats hang conspicuously in easily accessible roosts, they are more prone to disturbance by man than the Vespertilionidae (Tupinier, pers. comm.).

CONSERVATION MEASURES
Europe: It is protected in all countries. In Britain 4 of the 6 major breeding roosts (all buildings) have been given special protection and one was rebuilt just to accommodate the bats.

Caves in several countries have also been protected by legal agreement and gated to prevent unauthorized disturbance (Fairon, pers. comm.). All breeding roosts need protecting as well as a large range of hibernation roosts which seem vital for the species' survival. Individuals from one colony were known to live in areas up to 2500 km^2 (205).

ADDITIONAL BIBLIOGRAPHY
2, 14, 17, 25, 32, 33, 35, 49, 50, 54, 55, 57, 58, 59, 62, 66, 73, 77, 79, 80, 84, 86, 87, 95, 112, 126, 127, 130, 143, 145, 166, 172, 209, 211, 218, 224, 227, 232, 239.

CHIROPTERA: RHINOLOPHIDAE   Status: E   E

2. *Rhinolophus hipposideros* - Bechstein 1800

E. Lesser horseshoe bat
F. Petit rhinolophe fer à cheval
G. Klein Hufeisennase
I. Ferro di Cavallo minore

D. Dvaerghesteskonaese
N. Kleine hoefijzerneus
S. Murciélago pequeno de herradura
H. Rhinolophos i mikra

DISTRIBUTION

EEC: W Ireland, Wales and W and SW England, France, SE Belgium, Luxembourg, S Netherlands, S Germany, Italy, Sicily, Sardinia, Corsica and Greece.

Europe: Throughout Czechoslovakia, Switzerland, Austria, Hungary, Yugoslavia, Albania, Bulgaria, Rumania, Spain and Portugal. Also southern East Germany.

Poland: Localized to 3 areas, the largest population being in the south around Krakow, and also near Prizemysl and Nysa (Bogdanowicz, pers. comm. 1982, Bogdanowicz & Ruprecht, pers. comm.).

World: From Ireland, Iberia and Morocco through S Europe and N Africa to Turkestan and Kashmir.

HABITAT

Found in woodland, scrubland and grassland often near water. Hibernates in caves, mines and cellars mostly close to nursery sites, and usually forms breeding clusters in buildings in summer. Often cellars and roofs of the same building are used in winter and summer respectively.

POPULATION

EEC: This species has disappeared from northern areas of its range since 1950, and colonies appear to be declining (often up to 90%) over the entire range.

Britain: Observation of the area of guano piles of 10 nursery colonies in Devon and S Wales suggests that they have

Rhinolophus hipposideros

declined recently by over 50% but no cause is known (Stebbings' data).

Belgium: It is now limited to the southern part of Belgium, and numbers are greatly reduced. A survey from 1959 to 1977 showed that this species completely disappeared from a former population of 325, the last record being in October 1973 (69, 73, 93).

France: This is one of the most commonly encountered species but its decline is very marked, especially in the north. In winter it is found hibernating underground singly or occasionally in small groups, although some colonies remain in buildings over winter, eg a colony found at Arvillard which numbered several hundred bats.

Between 1936 and 1960, 120 were ringed in the Isère department — representing 1·4% of all *Rhinolophus hipposideros* ringing in France (157, 219, Tupinier, pers. comm.).

Germany: The species is almost completely extinct (Roer, pers. comm. 6/1982). A population numbering many hundreds in 1958 was halved by the mid-1960s, and was virtually gone by 1970 (244). Colonies once numbering hundreds are now extinct or down to a few tens.

Netherlands: It is now extinct in Dutch caves where up to 300 used to hibernate in the early 1940s; 80% of the decline had occurred by 1955 and most of the remainder by 1970 (233).

Europe: **Poland:** In the Raclawicka cave, a hibernating population was monitored over 30 years with the following results (236).

Year	Number
1950	300
1968	10
1971	3
1979	2

The species has declined substantially in 35 years to 1979 in caves of the Krakow-Czestochowa Uppland (114).

Rhinolophus hipposideros - lesser horseshoe bat

Czechoslovakia: In favourable areas, summer colonies of females number between 10 and 100, and hibernating colonies of both sexes total up to 300. However, in Bohemia, a large decrease in population has occurred recently. In NE Bohemia and Moravia, the decrease is relatively less, being 10-50%. Generally, it is one of the most threatened bat species (182).

Spain: A marked decline has occurred in Spain, but it is found in the majority of provinces. Two breeding colonies occur at Arredondo and Linares de Riofrio (221, Tupinier, pers. comm.).

World: No population trends are known elsewhere.

THREATS

There is no known cause of most of the observed declines, but little work has been done on this species. Climatic change is not thought to be responsible but rather habitat or land use changes, and perhaps organochlorine pesticides. As with *Rhinolophus ferrumequinum*, this species is found in conspicuous roosts during the summer and so is highly susceptible to human disturbance. Disturbance of bats in hibernacula and reductions in available food are also major causes of decline (182, 69, Tupinier, pers. comm.). In Poland, the main cause of the disappearance of the hibernating bats in the Raclawicka cave was thought to be the frequent speleological expeditions causing sustained disturbance to this very sensitive species (Bogdanowicz, pers. comm. 1981).

It has also been suggested that fox and badger gassing may be killing roosting bats in France (Noblet, pers. comm. 4/1981).

CONSERVATION MEASURES
Europe: Protected in all countries. A few colonies have been specially protected in several countries. Research is urgently required to ascertain causes of declines and to find solutions.

ADDITIONAL BIBLIOGRAPHY
2, 5, 14, 17, 23, 25, 27, 31, 33, 35, 43, 49, 57, 59, 62, 76, 78, 77, 79, 80, 81, 83, 84, 86, 87, 95, 112, 120, 126, 129, 130, 143, 145, 166, 168, 170, 172, 192, 196, 209, 211, 224, 232, 239.

3. *Rhinolophus blasii* - Peters 1866

E. Blasius' horseshoe bat
F. Rhinolophe de Blasius
G. Blasius-Hufeisennase

I. Rinolofo de Blasius
N. Blasius' hoefijzerneus
S. Murciélago dalmata de herradura

DISTRIBUTION

EEC: NW Italy (though not recorded in Marche and Emil Romagna regions (Vernier, pers. comm. 1981)), Sicily, Greece.

Europe: Yugoslavia, Albania, Bulgaria (Beron, pers. comm.).

World: From Italy through SE Europe and SW Asia to E Afghanistan, north to the Kopet Dag and Caucasus, Morocco and Tunisia.

HABITAT

Little known. Breeds in clusters in caves and mines.

POPULATION

EEC: Rarely found, and very little known.

Europe: **Bulgaria:** Relatively common in Bulgarian caves (Beron, pers. comm.).

World: In Israel it is on the verge of extinction (152).

THREATS

Collection, disturbance and loss of caves and habitats, and in Israel fumigation of caves with organochlorine pesticide (152) are main threats. This fumigation aims to kill the fruit-eating bat *Rousettus aegyptiacus.*

CONSERVATION MEASURES

Europe: Protected in all countries where it occurs.

ADDITIONAL BIBLIOGRAPHY

25, 61, 62, 145, 146, 173, Tupinier, pers. comm.

Rhinolophus blasii

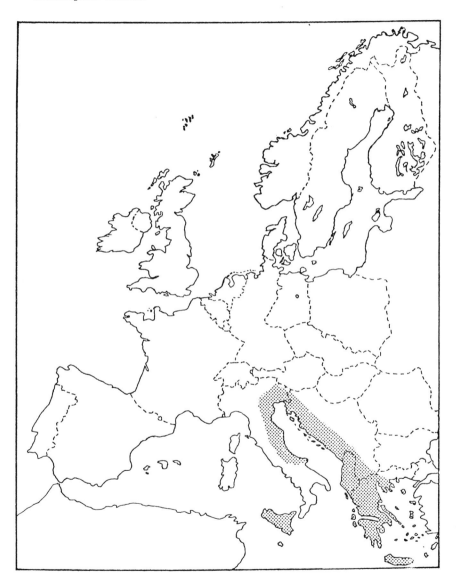

4. *Rhinolophus euryale* - Blasius 1853

E. Mediterranean horseshoe bat
F. Rhinolophe euryale
G. Mittelmeer-Hufeisennase
I. Rinolofo euriale

N. Paarse hoefijzerneus
S. Murciélago mediterraneo de
   herradura
H. Rinolophos i mesogiaki

DISTRIBUTION

EEC: Central and south France, Italy, Sicily, Sardinia, Corsica and
            Greece.

Europe: SE Slovakia only (182), Hungary, Yugoslavia, Albania, Austria,
            W Rumania, Bulgaria (Beron, pers. comm.), N Iberian penin-
            sula.

World: Mediterranean zone of S Europe and N Africa and east to
            Turkestan and Iran.

HABITAT

Little known. Mostly found in wooded country and generally
roosts in caves.

POPULATION

EEC: Little known. Declines noted in breeding colonies but little
            exact information available.

            France: Only 2 records in the Isère department, one ringed in
            1958, and one captured in February 1971 in Balme-les-Grottes
            (157).

            Uncommon in Rhône-Alpes region; in the Ardèche used to
            form important colonies, eg in the Grotte de Saint-Marcel more
            than 1000 individuals (219). Extinct in Ain but breeds in small
            numbers in the Jura (Rolandez, pers. comm. 5/1981). Howev-
            er, Tupinier (pers. comm.) suggests this species is extinct in
            France.

Europe: **Czechoslovakia:** Recent large declines in populations are
            reported, although some colonies of 100–200 remain. The total

Rhinolophus euryale

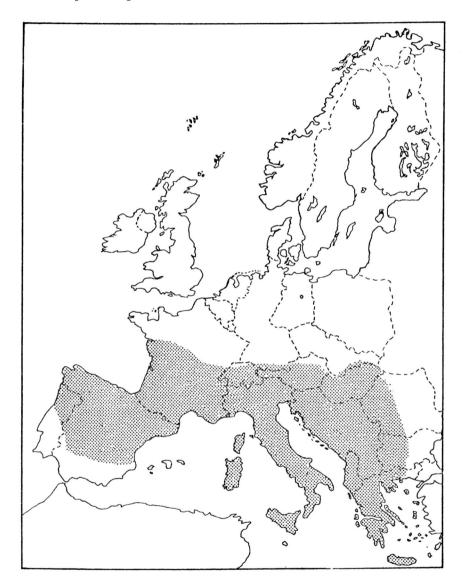

number of bats remaining in SE Slovakia is about 1000. It is considered the most threatened bat species (182).

Spain: A serious decline is occurring, though no figures are available (Tupinier, pers. comm.).

Switzerland: Used to occur but lack of any recent records suggests it is extinct (Aellen, pers. comm., Tupinier, pers. comm.).

World: No information is available.

THREATS

Disturbance in caves and loss of habitat, and fumigation of caves with pesticide (152).

CONSERVATION MEASURES
Europe: Protected in all countries where it occurs. Research is required to assess status and find conservation measures. Total protection of all cave roosts is urgently required. Some are already gated.

ADDITIONAL BIBLIOGRAPHY
14, 17, 25, 50, 54, 57, 56, 59, 62, 84, 87, 112, 126, 127, 145, 218, Vernier, pers. comm. 1981.

5. *Rhinolophus mehelyi* - Matschie 1901

E. Mehely's horseshoe bat
F. Rhinolophe de Mehely
G. Mehely-Hufeisennase
I. Rinolofo di Mehely

N. Mehely's hoefijzerneus
S. Murciélago mediano de herradura
H. Rinolophos i mehelios

DISTRIBUTION

EEC: Mediterranean coast of France (Tupinier, pers. comm.), SE Italy, Sicily, Corsica, Sardinia and Greece.

Europe: Bulgaria, Rumania, Spain, Portugal, Yugoslavia (Beron, pers. comm., 43, 221).

World: Known from few localities in S Europe, the Mediterranean islands, NW Africa, Asia Minor, the Caucasus, and the Zagreb Mountains, Iran.

HABITAT

Little known. Found in caves in summer and winter.

POPULATION

EEC: Small isolated colonies. Perhaps a relict species.

France: There are no recent records of this species. It has definitely disappeared in the south-west (221) and is practically extinct throughout (Noblet, pers. comm. 6/1983).

Europe: **Rumania:** This species has been recorded in the Rumanian Dobrogea both in summer and winter (63, 64). In 1974, a large nursery colony of 500 was found in the Rumanian Dobrogea, but in summer 1979 a colony of only 100–150 was recorded at the same locality.

World: No information.

THREATS

Disturbance, collecting and habitat loss are probably the main threats, but little is known.

Rhinolophus mehelyi

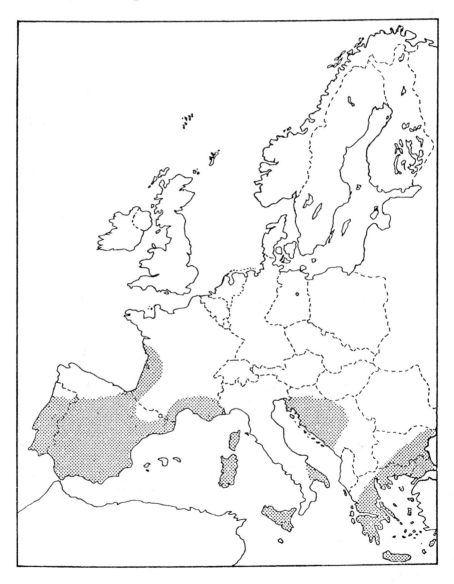

CONSERVATION MEASURES

Protected in all countries where it occurs. Research into habitat and roost requirements is needed urgently.

ADDITIONAL BIBLIOGRAPHY

7, 145.

6. *Myotis bechsteinii* - Kuhl 1818

E. Bechstein's bat
F. Vespertilion de Bechstein
G. Bechstein-Fledermaus
I. Vespertilio di Bechstein

N. Langoorvleermuis
S. Murciélago de Bechstein
H. Nycteris i Bechstein

DISTRIBUTION

EEC: Central S England, France, SE Belgium and Netherlands, Luxembourg, Germany, Italy (except south), southern tip of Sicily, Corsica (22, 148, Noblet, pers. comm. 6/1983, Tupinier, pers. comm.).

Italy: Records are scattered throughout the country except in extreme south and north-east (Vernier, pers. comm. 1981).

Europe: Bulgaria, Rumania, Hungary, Yugoslavia, Czechoslovakia, Austria, Switzerland, E Germany, S Sweden, Spain, Portugal and Poland (29, Bogdanowicz, pers. comm. 1982).

World: Europe from Spain and France, England, S Sweden east to W Russia, the Caucasus.

HABITAT

Woodland, woodland edge and parkland. Usually roosts in hollow trees in summer and additionally found in caves, tunnels and mines in winter. Feeds by gleaning insects from foliage and around and amongst trees. Now found and breeds in artificial roost boxes, particularly in Germany and Czechoslovakia.

POPULATION

EEC: Generally very rare, although there is some archaeological evidence that it was much more abundant 2000+ years ago (Stebbings' data, Rolandez, pers. comm. 5/1981, 182). It forms small colonies in summer, but it is usually solitary in hibernation. Probably still declining and may be endangered throughout range.

Myotis bechsteinii

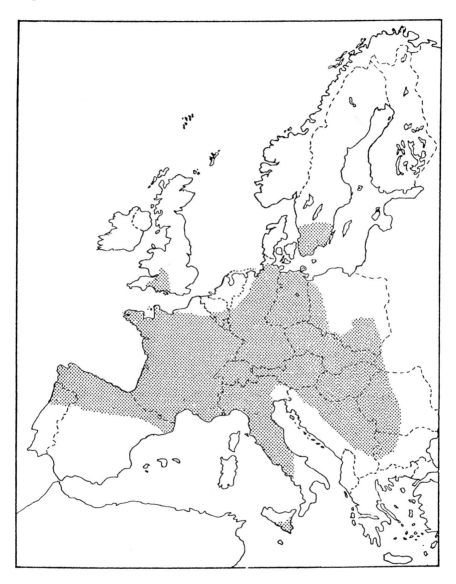

Myotis bechsteinii - Bechstein's bat

Britain: Very rare, there are only a small number of authenti-cated records for this species (202).

France: Very few records and is regarded as very rare. It has been ringed in Ardèche and the Rhône (157, Tupinier, pers. comm., 219). This species has been identified in Corsica from skulls in owl pellets (22, 186, 148, Tupinier, pers. comm.).

Germany: Small populations occur throughout (Roer, pers. comm.).

Netherlands: First recorded in 1938 when 2 bats were found in South Limburg. There are a number of isolated records from the mines of S Limburg but none recent (23, Glas, pers. comm.).

Europe: **Bulgaria:** Rare, but found throughout to Black Sea coast (Beron, pers. comm.).

Czechoslovakia: Found infrequently throughout, this species is rare and usually only isolated individuals are observed (182).

Hungary: Minute population (Topál, pers. comm.).

Iberian peninsula: There are only 5 records from Spain and Portugal.

World: No information.

THREATS

Loss of roosts and also particularly of high forest are probably the most important causes of declines, but little known.

CONSERVATION MEASURES

Europe: Protected in all countries. Research is required to critically establish habitat and food requirements, but because of its extreme rarity this would be difficult.

ADDITIONAL BIBLIOGRAPHY

2, 16, 24, 32, 33, 35, 49, 58, 73, 77, 79, 80, 81, 83, 84, 87, 91, 103, 120, 126, 129, 133, 143, 164, 166, 190, 191, 192, 209, 211, 232, 235, 239.

CHIROPTERA: VESPERTILIONIDAE   Status: EEC World
                                                V   V

7. *Myotis nattereri* - Kuhl 1818

E. Natterer's bat
F. Vespertilion de Natterer
G. Fransen Fledermaus
I. Vespertilio di Natterer

D. Frynseflagermus
N. Franjestaart
S. Murciélago de Natterer
H. Nycteris i Krossoti

DISTRIBUTION

EEC: Throughout France, Corsica, Belgium, Great Britain, Ireland,
        Italy, Sicily, Luxembourg, south Netherlands, Germany (except
        NW), Denmark (south of the Limfjord and Bornholm) (Baagøe,
        pers. comm. 1981).

Europe: Bulgaria (Beron, pers. comm.), Czechoslovakia, Austria,
        Switzerland, Hungary, Poland, East Gerrnany, Spain, Portugal,
        S Sweden, Norway (near Oslo) (Pedersen, pers. comm.
        4/1981), SE Finland (Kotka area and Imatra) (193).

World: Europe (except south-east and most of northern Scandinavia)
        but including S Norway and Sweden, Britain and Ireland,
        Morocco, Crimea, Caucasus, Palestine, Kopet Dag, Tadzhikis-
        tan, SE Siberia, Korea, Kyushu and Honshu.

HABITAT

Lives in forest, parkland and urban areas. Feeds along forest
edges and around trees, often close to water. Roosts in hollow
trees and buildings in summer, and additionally caves, mines
and tunnels in winter.

POPULATION

EEC: Known to have declined in some areas, but generally little
        known. Most colonies are small, up to 30, but occasionally
        forms colonies up to a few hundred, mostly in buildings.
        Appears to be rare over large areas of Europe, especially in the
        south.

British Isles: Widespread but more abundant in the south and
in Ireland (202).

Myotis nattereri

Myotis nattereri - Natterer's bat

Denmark: Of about 1200 bats hibernating in 2 caves in Jutland in 1981–82, only 7% were *M. nattereri*.

France: The species is unknown in Isère, except from barn owl pellets (157). Generally rare in SE France but there are some hibernating populations in mines at Beaujolais (219). It was recorded in Corsica in 1973 (186) and 1981 (148), in remains in owl pellets (Noblet, pers. comm. 6/1983, 22).

Italy: Apparently there are no records from the east coast.

Netherlands: Hibernating populations in Limburg declined from about 70 bats in the early 1940s to 10 in the late 1960s and 1970s (233). In South Limburg caves 1550 bats were banded in 10 years (28).

Europe:
Czechoslovakia: Relatively rare except in the south Bohemian basin and the Piedmont of Sumava Mountains. Summer colonies number 10–30, but in winter they usually occur individually (182). In 10 years, 19 nursery colonies totalling 368 specimens were examined; average 19 per colony (44).

Hungary: Very small populations (Topál, pers. comm.).

Poland: Populations in Koralowa cave in the Krakow-Czestochowa Uppland declined substantially in 35 years (114, 140).

Iberian peninsula: It has been recorded occasionally throughout the peninsula but only one colony is recorded (241, 242).

World: No information.

THREATS

Known to be killed in remedial timber treatment, but populations have also probably been reduced by loss of hollow trees, caves and mines.

CONSERVATION MEASURES

Europe: Protected in all countries. One of the most widespread species, but little is known about its habitat or food requirements. Hibernation sites in several countries have been protected.

ADDITIONAL BIBLIOGRAPHY

1, 2, 5, 11, 12, 13, 32, 35, 49, 58, 65, 73, 77, 79, 80, 81, 83, 84, 87, 91, 101, 103, 108, 115, 120, 126, 129, 143, 145, 164, 166, 185, 190, 191, 192, 196, 209, 211, 224, 232, 239.

CHIROPTERA: VESPERTILIONIDAE   Status: V(?E)   K(?E)

8. *Myotis capaccinii* - Bonaparte 1837

E. Long-fingered bat
F. Vespertilion de Capaccini
G. Langfuszfledermaus

I. Vespertilio de Capaccini
S. Murciélago patudo
H. Nycteris i macropus

DISTRIBUTION
EEC: SE France, Italy, Sicily, Corsica, Sardinia, Greece (157, Rolan-
           dez, pers. comm. 5/1981, 219, Noblet, pers. comm. 4/1981,
           Vernier, pers. comm. 1981, Aellen, pers. comm.).

           Greece: Present on Thassos and Crete (Beron, pers. comm.
           145).

Europe: Bulgaria (Beron, pers. comm.), Yugoslavia, Spain (221).

World: Mediterranean zone of Europe and NW Africa, including most
           of the Mediterranean islands, S Asia Minor and Palestine, S
           Iraq, S Iran lower Amu-Darva, Uzbekistan.

HABITAT
           Riparian woodland and scrub habitats. Usually roosts in caves
           and mines, both in summer and winter.

POPULATION
EEC: Little known. Forms large colonies, sometimes mixed with
           other species.

           France: Has disappeared in parts of south-east France, the
           northern limit now being a strictly Mediterranean area (157,
           Noblet, pers. comm. 4/1981, Rolandez, pers. comm. 5/1981,
           219).

Europe: **Spain:** Very few known captures are all limited to the
           Mediterranean zone. Often lives in multi-species colonies
           where *Miniopterus schreibersii* dominates, *Myotis capaccinii*
           only representing 0.5% of the mixed population (15, 221).

World: Nothing known.

THREATS
           Vulnerable to disturbance and loss of cave roosts. It has been
           collected, and killed by vandals.

Myotis capaccinii

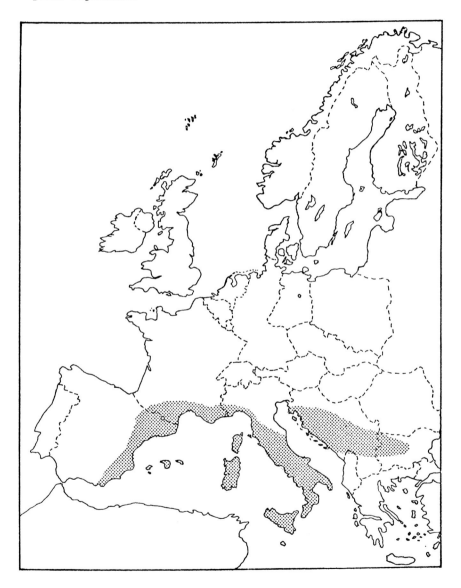

CONSERVATION MEASURES

Protected in all countries. Research urgently required to assess status and habitat requirements. Known colonies need adequate protection from disturbance.

ADDITIONAL BIBLIOGRAPHY

2, 24, 25, 50, 59, 62, 146.

EEC World
Status: E E

9. *Myotis dasycneme* - Boie 1825

E. Pond bat
F. Vespertilion des marais
G. Teichfledermaus
I. Vespertilio dasicneme

D. Damflagermus
N. Meervleermuis
S. Murciélago de los marismas
H. Nycteris i limnobios

DISTRIBUTION

EEC: Germany, Belgium, Netherlands, Denmark, probably Luxembourg (but not yet recorded), extreme NE France.

Netherlands: Present in Friesland and north Holland provinces. Also in south-west of Holland (Glas, pers. comm., Voûte, pers. comm. 3/1982).

Denmark: Found in Jutland but no recent records elsewhere (Baagøe, pers. comm. 1981).

Europe: Sweden, East Germany, Czechoslovakia, Poland, NE Hungary, N Rumania, Bulgaria.

Sweden: Recently recorded in south Sweden (in Uppland, north of Stockholm) (Gerell, pers. comm. 1981).

Poland: Scattered records throughout the country (Bogdanowicz & Ruprecht, pers. comm.).

Czechoslovakia: Mainly in west, N and NW Bohemia, Moldavia and east Slovakian lowland (182).

World: NE Europe and W Siberia, NE France and Germany, east to River Yenisey. It is found between latitudes 48–60°N. In eastern Europe and western Siberia, the scarcity of records makes it difficult to assess numbers and distribution. The species is known to migrate (Horáček, pers. comm., Sluiter *et al.* 1972, via Horáček, pers. comm.).

HABITAT

Prefers riparian habitats and usually feeds over waterways (197). It roosts in buildings in summer and caves, mines and cellars in winter. Nursery roosts and hibernacula are often 200–300 km apart.

Myotis dasycneme

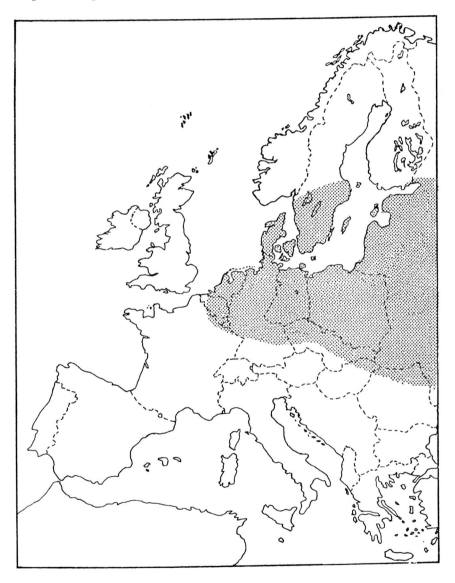

POPULATION

EEC: It appears that there are 2 centres of population in western Europe, in Denmark and the Netherlands. Although exact numbers are unknown, total numbers in western Europe may be less than 3000 bats (Baagøe, pers. comm. 1982, Voûte, pers. comm. 3/1982). Bats from the remaining nursery roosts in NW Netherlands hibernate in SE Netherlands, N France, Belgium and Germany. Some may travel further east. In the Netherlands many nursery roosts have been lost and large declines have occurred (231).

Netherlands: In the Friesland province 3 nursery colonies are known in churches.

Tjerkwerd	— colony of about 400 adult females
Berlikum	— colony of about 170 adult females
Oosterlittens	— colony of just a few adult females

and in north Holland:

Kwadijk	— colony of about 150 adult females
Wieringerwaard	— colony of about 150 adult females

Assuming equal numbers of males, in summer about 1700–2000 adults are present in the Netherlands (Voûte, pers. comm. 2/1982). Marking individuals has shown bats migrate between adjacent countries (94).

Denmark: Found only in the chalk mines in Jutland (Baagøe, pers. comm. 1981 and 1982). This species probably breeds in mid-Jutland. In 1981–82, up to 800 bats were found in 3 mines, Smidie, Tingbaek and Mønstead (12, Jensen, pers. comm.).

Germany: Occasional visitor for hibernation. There are no known breeding colonies (Roer, pers. comm.).

Europe: **Czechoslovakia:** Previously recorded in 10 localities mostly in winter, with the exception of the Tisza basin where it was found during the summer of 1980 (Horáček, pers. comm.). Recently found in small numbers in Moldavia (226, Horáček, pers. comm.).

Hungary: Occasional records with probably no breeding colonies (Topál, pers. comm.).

Myotis dasycneme - pond bat

World: The species is very rare in central Europe. Other population centres are known in the USSR.

USSR: In 1969, over 1000 hibernating bats recorded in the Smolinska cave near Kamensk-Uralsk (211), but in 1974 only a few individuals were found (212). There is also a decline in the other mass hibernacula in the artificial cave 'Sablinskye pestchery' near Leningrad. Other breeding areas include the basins of the Rivers Volga and Don, and in White Russia — where it is considered rare (Kurskov, pers. comm., Horáček, pers. comm.). The species may total fewer than 7000 bats worldwide and it should be considered as highly endangered.

THREATS

Major declines have been caused by remedial timber treatment in buildings containing nursery colonies but collecting, disturbance while in hibernation and loss of roosts have all contributed. Pollution of waterways reduces food.

CONSERVATION MEASURES

Europe: Protected in all countries where it occurs. Control of water pollution is required, as well as special protection being afforded to every breeding roost and major hibernation site.

The most important known roosts are in the Netherlands and Denmark (231, 12). Some sites are already protected, especially in the Netherlands (Vernier, pers. comm. 1981). Smidie cave in Denmark was protected in 1978 (Jensen, pers. comm.).

ADDITIONAL BIBLIOGRAPHY
11, 13, 24, 28, 32, 33, 34, 36, 49, 58, 65, 73, 84, 86, 91, 95, 96, 115, 137, 143, 164, 166, 168, 172, 190, 191, 192, 196, 209, 232.

10. *Myotis daubentonii* - Kuhl 1819

E. Daubenton's bat
F. Vespertilion de Daubenton
G. Wasserfledermaus
I. Vespertilio di Daubenton

D. Vandflagermus
N. Watervleermuis
S. Murciélago de ribera
H. Nycteris i hydrobios

DISTRIBUTION

EEC: Throughout, except possibly the extreme N of Scotland, and Greece.

Europe: Spain, Portugal, Austria, Switzerland, Czechoslovakia, E Germany, Poland, S Scandinavia, Hungary, N Yugoslavia, Bulgaria and possibly Rumania.

Finland: South of line from Jakobstad to border near Lieksa (193).

World: Europe and S Siberia, east to Vladivostok, Korea and Manchuria, Britain and Ireland, S Scandinavia, Kurile Islands, Sakhalin and Hokkaido.

HABITAT

Found primarily in riparian habitats and around ponds and lakes, often roosting under rock ledges or in scree, as well as in cellars, mines, caves, under bridges, in tunnels, in buildings and hollow trees. Feeds over water often on emerging insects. Summer and winter roosts may be widely separated.

POPULATION

EEC: Appears moderately common throughout Europe. Perhaps one of the most abundant of all species. Populations are generally stable but some declines have been recorded in long-known nursery roosts (Stebbings' data). Some increases have also been recorded in hibernacula in the last 20 years, but it is not known whether these are due to a real increase in abundance of the species or greater congregation in fewer available roosts (49, Stebbings' data).

Britain: Widespread north to Inverness and probably including all mainland Scotland. In summer the females often form large nursery colonies of 100+ females in buildings, tunnels and bridges, but are usually solitary during the winter.

Myotis daubentonii

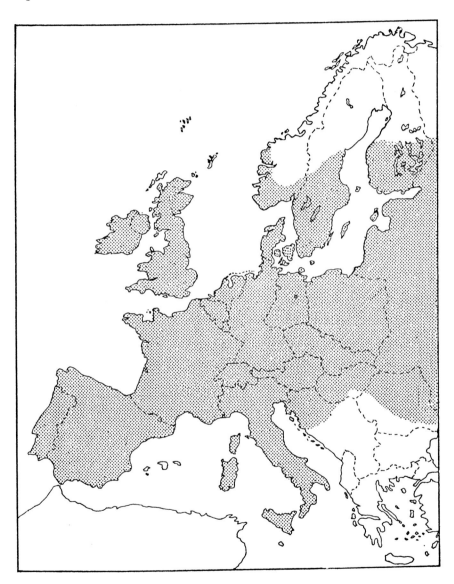

Myotis daubentonii - Daubenton's bat

Denmark: Appears to be increasing in Denmark, but the cause is unclear (49). Smidie cave in Jutland had about 525 *Myotis daubentonii* out of 700 counted in hibernation in the winter of 1981–82. Likewise it dominates in the 2 Tingbaek caves and at Alborg (Jensen, pers. comm.). The largest population of about 3000 is found in the Mønstead cave (12).

France: Little is known of this species in France, but it is probably better represented than the few records suggest (157, Noblet, pers. comm. 4/1981, Rolandez, pers. comm. 5/1981, 219).

Netherlands: The most abundant species in hibernation in South Limburg. A population in 12 mines increased from about 30 bats to 150 between the mid-1940s and the late 1970s (233).

Europe: **Czechoslovakia:** In Bohemia the summer colonies generally number 20–100 bats, but in winter they occur singly. In Moravia they are infrequent, and in Slovakia rare. This is the only bat species showing a slight increase (182).

Iberian peninsula: Rarely observed, but is thought to occur throughout. Recent records come from 2 regions: firstly, Cabezarubias where almost 100 male bats were found distributed in small groups up to 12 in a tunnel; secondly, 2 localities in the Cantabrique Mountains, Ramales and Arredondo, where the bats were found in cracks in bridges in groups of 2 or 3 (221).

Rumania: Apparently very rare with only 6 records, 2 from the Black Sea coast (44, 125).

Sweden: Very abundant in south Sweden (Gerell, pers. comm. 1981).

World: As above.

THREATS

The main threats are disturbance in hibernation and loss of roosts, as well as pollution in rivers, lakes, etc. Bats have been killed by vandals, especially while hibernating. Nursery roosts in buildings are threatened by remedial timber treatments.

CONSERVATION MEASURES

Protected in all countries. Many caves have been protected in several European countries in which this is the dominant species. Studies are required to assess status in most countries and to define habitat requirements. Is the observed increase of hibernating populations in north-west Europe a true pattern overall, or is it due to bats crowding into fewer available roosts?

ADDITIONAL BIBLIOGRAPHY

2, 5, 11, 13, 24, 25, 26, 28, 32, 33, 36, 51, 58, 65, 73, 77, 81, 83, 84, 86, 87, 91, 95, 101, 103, 108, 115, 120, 126, 129, 137, 138, 143, 160, 164, 166, 172, 181, 190, 191, 192, 196, 209, 211, 232, 235, 239.

11. *Myotis nathalinae* - Tupinier 1977

E. Lesser Daubenton's bat
F. Petit vespertilion de Daubentoni (Murin Nathaline)
G. Kleine Wasserfledermaus
Described as new to science in 1977 (222) but differentiation remains unclear.

DISTRIBUTION

EEC: France and may occur in other countries.

France: Isère and Indre departments, also in Corsica (Noblet, pers. comm. 4/1981, 6/1983).

Europe: Recorded from Spain, Switzerland, Poland, but distribution poorly known.

Poland: Includes Pomeranian lake region, Wielkopolska-Kujawy and Masovian lowlands, Malopolska upland, Bialoweiza and the Lubuskie lake district (30, 178).

Spain: Ciudad-Real province, Salamanca region, west Pyrenees (23, Tupinier, pers. comm., 131).

World: As above.

HABITAT

Unknown. Probably similar to *Myotis daubentonii,* living in riparian habitats and roosting in buildings, tunnels, caves and trees. Probably feeds over water.

POPULATION

EEC: Unknown.

Europe: Unknown.

World: Unknown.

Myotis nathalinae

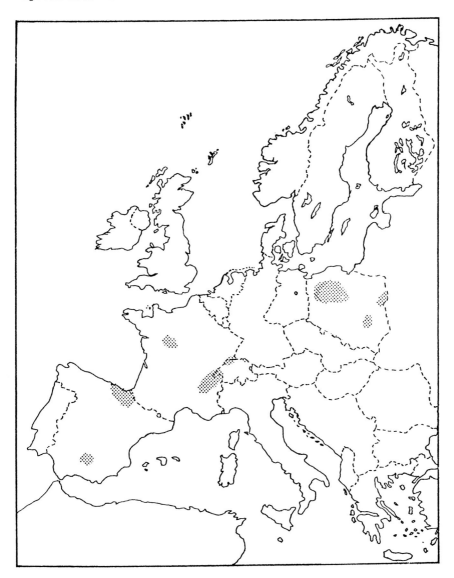

THREATS

Presumably similar to *M. daubentonii*. Water pollution and pesticides are likely to be the greatest threats, as well as loss of roosts.

CONSERVATION MEASURES

Protected in all countries where it has been found and will be protected by current legislation in most other countries if a wider distribution is found.

ADDITIONAL BIBLIOGRAPHY

5, 222, 239.

12. *Myotis emarginatus* - Geoffroy 1806

E. Notch-eared bat
F. Vespertilion à oreilles echan-
crées
G. Wimperfledermaus
I. Vespertilio emarginato

N. Ingekorven vleermuis
S. Murciélago de Geoffroy
H. Nycteris i blephardoti

DISTRIBUTION

EEC: S Germany, Luxembourg, Belgium, S Netherlands, all France except possibly the north-west, Italy, Sicily, Corsica, Sardinia, Greece.

Europe: Austria, Switzerland, Hungary, Czechoslovakia, Yugoslavia, Albania, Rumania, Bulgaria.

Spain: Ebre a Pons basin, Huesca, Soria, Santander and Bilbao regions.

Czechoslovakia: Mainly in southern and central Moravia and Slovakia, scattered in Bohemia.

World: Southern Europe (north to Netherlands and Czechoslovakia), Crimea, Caucasus, Kopet Dag, east to Tashkent and E Iran, Palestine and Morocco. Used to occur in S Poland but not seen there since 1962 (Bogdanowicz, pers. comm. 1981).

HABITAT

Feeds in woodland, parkland, scrubland and pasture. Nursery roosts are mainly in buildings and probably trees, as well as cellars and caves. Hibernates in caves, mines and cellars in relatively warm places between 4.5 and 9°C (82).

POPULATION

EEC: Rare and patchily distributed in northern areas but locally forms large colonies in caves in the south.

Belgium: Very rare in low-lying districts – ie northern parts where only 2 nursery colonies are known (Jooris, pers.

Myotis emarginatus

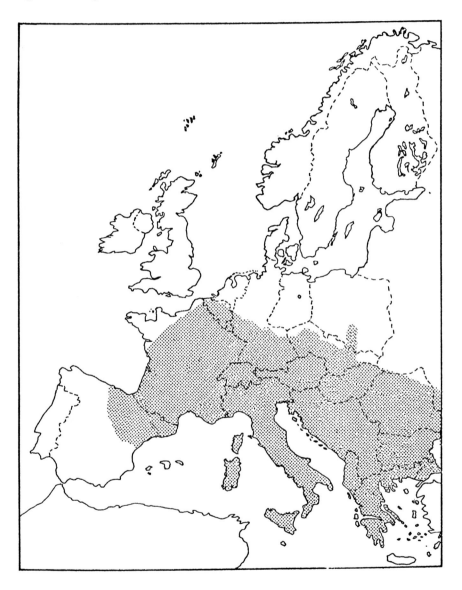

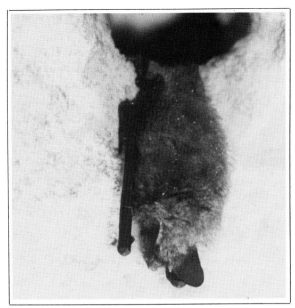

Myotis emarginatus - notch-eared bat

comm.). In the summer of 1980 a colony numbering about 100 bats was found in Durbuy.

France: Population declines have been reported together with horseshoe bats, with which this species is often associated.

Isère department – hibernating individuals are in disused quarries in the Rhône valley.

In Ardèche a nursery colony of 1000 bats was found, but records are mainly of isolated individuals. Ain region has some small breeding colonies.

Italy: No recent records from the south-east coast.

Netherlands: During a 15-year survey in South Limburg mines from 1936 to 1951, this was one of the most abundant hibernating species. Between 1942 and 1951, 2062 were banded (24). A nursery colony occurred in the mines.

Large declines occurred in the hibernating populations from about 200 in 1945 to only 20 in the 1970s (233).

Europe: Czechoslovakia: Generally very rare, with larger numbers being found in Moravia and Slovakia (182).

Poland: It is extinct in Poland (Bogdanowicz, pers. comm. 1981).

Rumania: A nursing colony in the Rumanian Dobrogea was present in 1955 and 1958, but since then only one adult male was found (44).

Spain: Recorded for the first time in 1964. Four breeding colonies have been found at Arrendondo, Cueva de la Abegas, Saja and San Leonardo. At all these localities *M. emarginatus* was found roosting with *Rhinolophus* spp. (243).

World: Unknown.

THREATS
Vulnerable to disturbance and loss of roosts, particularly those in caves.

CONSERVATION MEASURES
Protected in all countries. Some cave-roosting populations in several countries have been protected. Little is known of the ecology, critical habitat or status of this species.

ADDITIONAL BIBLIOGRAPHY
2, 5, 16, 25, 26, 28, 32, 36, 49, 57, 62, 73, 83, 84, 86, 87, 91, 126, 127, 143, 145, 166, 196, 209, 211, 218, 232, 239.

13. *Myotis mystacinus* - Kuhl 1819

E. Whiskered bat   D. Skaegflagermus
F. Vespertilion à moustaches   N. Baardvleermuis
G. Kleine Bartfledermaus   S. Murciélago bigotudo
I. Vespertilio mustaccino   H. Nycteris i mystakophoros

(Because of former confusion with *M. brandtii*, it is not yet clear what details refer just to this species.)

DISTRIBUTION
EEC: Probably throughout, except most of Denmark and northern Scotland. Only a few scattered records from most countries but will probably occur in the area shown on the map. Its apparent absence from Denmark (excepting Bornholm, (10)) is interesting because *M. mystacinus* and *M. brandtii* seem to be sympatric over the remainder of their range.

Europe: Throughout, but again only a few scattered records.

World: Entire Palaearctic from Ireland to Japan, north to about 65° in Europe and west Siberia, south to Morocco, north to Iran and the Himalayas.

HABITAT
Woodland edge, scrubland and riparian habitats. Roosts mostly in buildings but also occurs in hollow trees in summer, and additionally caves, mines and cellars in winter.

POPULATION
EEC: Little known, but forms small scattered colonies. It is a rare bat over much of Europe. Some populations are known to have declined, but others are possibly increasing slightly, eg Belgium (Fairon, pers. comm., 73).

BRITAIN: Positive records from southern half, but probably found throughout England and Wales. In SW Wales, one of the most abundant species forming nursery colonies in buildings.

Myotis mystacinus

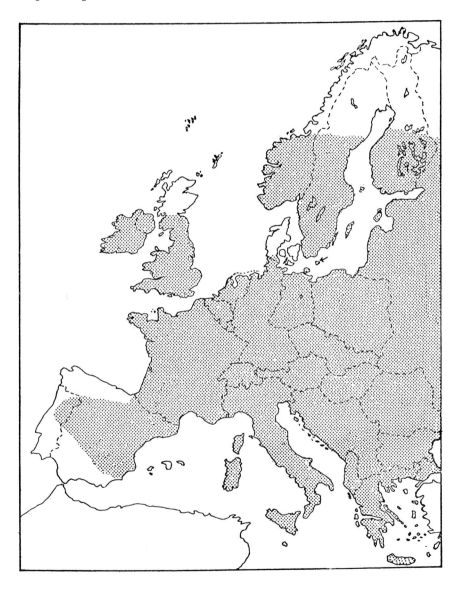

Belgium: A survey of the Grand Carrière de Romont from 1957 to 1980 showed *M. mystacinus* constituting about 30% of the hibernating bat population, and the proportion appears to be increasing (93).

France: Rarely found, with only 2 records from Isère, a few ringed in Ardèche, and it has been captured at Bretolet. Small numbers are found throughout the Ain region.

Does not appear threatened in France, but no detailed information on the population is available. It is found in bat boxes in the Lyonnaise region (157, Rolandez, pers. comm. 5/1981, 219, Tupinier, pers. comm.).

Italy: Patchily recorded in many areas but there are no observations for the south or east peninsula regions, although it is assumed to be present (Vernier, pers. comm. 1981).

Netherlands: An abundant species in hibernation in South Limburg mines. In a 15-year survey (1936–51) in South Limburg, 1378 bats were ringed, showing this species to be very common (23).

Europe: **Czechoslovakia:** Populations appear stable. A hibernating population in the Dobsinska Ice Cave (Slovakia) numbers 300 bats. Usually the nursery colonies contain only a few individuals but they can number up to 30 (182).

Hungary: Very rare (Topál, pers. comm.)

Norway: Stable population (Pedersen, pers. comm. 4/1981).

Poland: Widespread in cellars and caves, eg they have been found in 2 caves in the Krakow-Czestochowa Uppland – the Ciemna and Nietoperzowa – between 1954 and 1979.

Rumania: During a survey of the Rumanian Dobrogea in 1974 and 1979 this species was not found, but had been recorded until 1958. A hibernating colony was found in 1956 (44, 63, 64).

Spain: There are very few records for this species, but it appears to be on the decline and has disappeared from some regions, eg Valencia and Murcia (Tupinier, pers. comm., 249).

Myotis mystacinus - whiskered bat

Sweden: Not threatened, widespread and common (Gerell, pers. comm. 1981).

World: Nothing known.

THREATS

Disturbance in caves, loss of roosts and remedial timber treatment in buildings.

CONSERVATION MEASURES

Protected in all countries. In Poland, the Nietoperzowa and Ciemna caves in the Ojcowski National Park have been protected for this and other species.

ADDITIONAL BIBLIOGRAPHY

1, 2, 5, 11, 12, 13, 28, 32, 33, 34, 36, 49, 58, 62, 65, 77, 79, 80, 81, 83, 84, 86, 87, 88, 91, 92, 95, 101, 103, 110, 115, 120, 126, 129, 130, 137, 138, 143, 145, 156, 160, 166, 181, 190, 191, 192, 196, 209, 211, 224, 232, 235, 236, 239.

14. *Myotis brandtii* - Eversmann 1845

E. Brandt's bat
F. Vespertilion de Brandt
G. Groszbartfledermaus
I. Vespertilio di Brandt

D. Brandt's flagermus
N. Brandt's vleermuis
S. Murciélago de Brandt

(First recognized in 1971. Therefore details are little known.)

DISTRIBUTION

EEC: Throughout, except Ireland and Scotland. It may occur in
            Ireland but not yet found. Few records from most countries but
            will probably occur as shown on the map.

Europe: Much of Europe, Norway and Sweden north to 64°, east to
            Urals and south to Spain.

            Iberian peninsula: Not yet recorded (Tupinier, pers. comm.).

            Poland: Said not to occur in the north (Bogdanowicz, pers.
            comm. 1981).

World: Unknown.

HABITAT

Found around woods, agricultural and rural areas. Roosts in
buildings and hollow trees in summer and mostly caves and
mines in winter. Often roosts in cool moist areas.

POPULATION

EEC: Little known, but already found more or less throughout region
            together with *M. mystacinus*, with which it was confused.
            Mostly small colonies in summer but occasionally forms
            colonies up to a few hundred. Some known to be in decline. It
            appears to be a rare bat over much of Europe and more so than
            M. mystacinus.

            Britain: Found roosting touching *M. mystacinus* in hibernation
            in caves and mines, but generally occurs in small numbers.
            Nursery colonies in houses have up to 50 bats.

Myotis brandtii

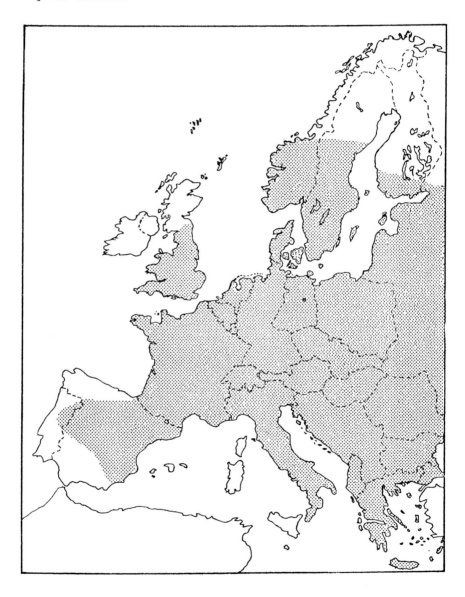

Myotis brandtii - Brandt's bat

Denmark: Smidie cave on the Jutland east coast had 50 *M. brandtii* out of a total of 700 hibernating bats counted in the winter 1981–82. Tingbaek caves had no *M. brandtii* out of 500 bats (Jensen, pers. comm.), but Mønstead cave had 50–70 bats of this species (*circa* 2% of hibernating population) (12).

France: Recorded at Chantilly in Aisne (223), and apparently very rare in France as a whole (Tupinier, pers. comm.).

Europe: **Hungary:** Apparently rare (Topál, pers. comm.).

Switzerland: Appears to have been found in 2 caves in the Jura Mountains (223).

World: Unknown.

THREATS

Colonies known to have declined substantially in Wales due to remedial timber treatment (McOwat, pers. comm.). Loss of hollow trees and caves and habitat modification are other major threats.

CONSERVATION MEASURES
Protected in all countries. Some populations have been protected by grilles and special reserves in a number of countries.

ADDITIONAL BIBLIOGRAPHY
10, 11, 34, 36, 49, 72, 73, 79, 80, 83, 84, 88, 99, 103, 110, 126, 129, 143, 153, 164, 166, 176, 179, 181, 189, 190, 216, 232, 236, 239.

15. *Myotis blythi* - Tomes 1857

E. Lesser mouse-eared bat
F. Petit murin
G. Kleines Mausohr

I. Vespertilio di Monticelli
H. Nycteris i oxygnathos

DISTRIBUTION - partly migratory
EEC: SE France, Italy, Sicily, Corsica, Sardinia and Greece. Does not occur in Germany (Roer, pers. comm.). A single specimen was found in England (Stebbings' data).

Europe: Austria, Switzerland, Czechoslovakia, Hungary, Yugoslavia, Albania, Bulgaria, Rumania, Iberian peninsula.

Czechoslovakia: Northern boundary of distribution; in summer occurs in south Slovakia only, in winter extends into central Slovakia and Moravia (182).

World: Whole of the Iberian peninsula (163) and Mediterranean zone of Europe and NW Africa, Crimea, Caucasus, Asia Minor, Palestine to Kirgizia, Afghanistan, Himalayas, Shensi and perhaps Inner Mongolia.

HABITAT
          Feeds in woods, parkland and on the edge of rural areas. In summer, roosts are mostly found in buildings and warm parts of caves but the species is also occasionally found in hollow trees. In winter it usually hibernates in caves but may occur in hollow trees.

POPULATION
EEC: Little known, but former, often large, colonies are now much reduced or absent.

France: In Ain colonies numbering several hundred bats have disappeared following disturbance of the colonies during cave excavations (Rolandez, pers. comm. 5/1981).

In the Isère department there were few records, but breeding colonies are known to be present. In 1966 a small colony of

Myotis blythi

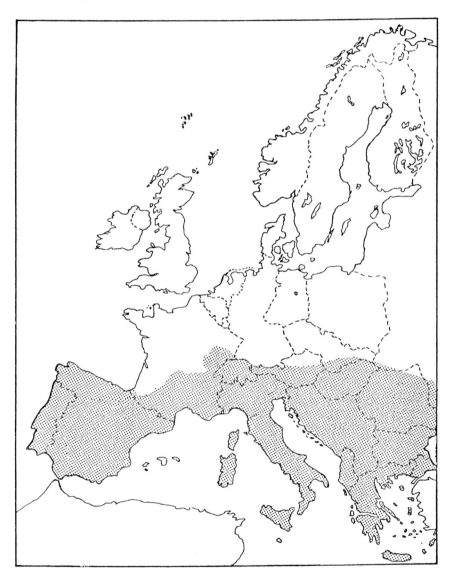

150–300 bats was found near Cremieu, with up to 40% of the bats being *Myotis myotis*. There are also records of this species in barn owl pellets (eg at Barraux) (157).

Nursery colonies are known in the Jura (Rolandez, pers. comm. 5/1981).

In Corsica *M. blythi* was found at Sagone and again in the Galeria region (23).

Europe: **Czechoslovakia:** In some areas of south Slovakia this is a relatively abundant species, with colonies numbering a few hundred. However, it is thought that a slight decrease in population is taking place (182).

Rumania: A large decline is reported in the Rumanian Dobrogea on the Black Sea coast. In one cave a nursery colony of 4000–5000 recorded in 1974 numbered only 150–200 by 1979 (44).

Switzerland: Apparently rare being recorded only in Valais, Tessin and Vaud (Aellen, pers. comm.).

World: In Israel it is on the verge of extinction (152).

THREATS

It is collected and disturbed in caves, killed by vandals, excluded from buildings, and killed during remedial timber treatment. Loss of large insects as a result of habitat changes is further threatening this large bat.

CONSERVATION MEASURES

Protected in all countries where it occurs. Nothing known about critical habitat. Research is urgently required.

ADDITIONAL BIBLIOGRAPHY

16, 17, 22, 26, 50, 59, 62, 83, 84, 86, 87, 111, 126, 143, 145, 146, 211, 224, 227, 239.

16. *Myotis myotis* - Borkhausen 1797

E. Mouse-eared bat
F. Grand murin
G. Grosz Mausohr
I. Vespertilio maggiore

N. Vale vleermuis
S. Murciélago ratonero grande
H. Nycteris i pontikootos

DISTRIBUTION - partly migratory
EEC: Not in Denmark or Ireland and is now essentially extinct in Britain, the Netherlands and northern Belgium, but it occurs throughout the rest of the region.

Europe: E Germany, Poland, Switzerland, Austria, Czechoslovakia, Hungary, Rumania, Bulgaria, Yugoslavia, Albania.

Poland: Found throughout except extreme NW and NE (Bogdanowicz & Ruprecht, pers. comm.).

World: Central and S Europe east to Ukraine, S England, most of the Mediterranean islands, Asia Minor, Lebanon and Palestine.

HABITAT
It is found in open woodland, parkland and in areas of old pasture. It is often associated with marginal urban areas, roosting in buildings and warm parts of caves and occasionally hollow trees in summer, and mainly caves, mines and cellars in winter. Summer and winter roosts may be widely separated.

POPULATION
EEC: Very large colonies were known totalling many hundreds or thousands of bats. Many of these throughout the region have declined substantially and the species is virtually extinct in northern areas.

Britain: Two colonies were known. The first was found in the 1950s but it had died out within 10 years, probably mainly due to disturbance and collecting. The second colony was discovered in 1969 but it appeared that the nursery cluster was killed in 1974. Its whereabouts was unknown. Only one male is now known to survive (Stebbings' data, January 1986).

Myotis myotis

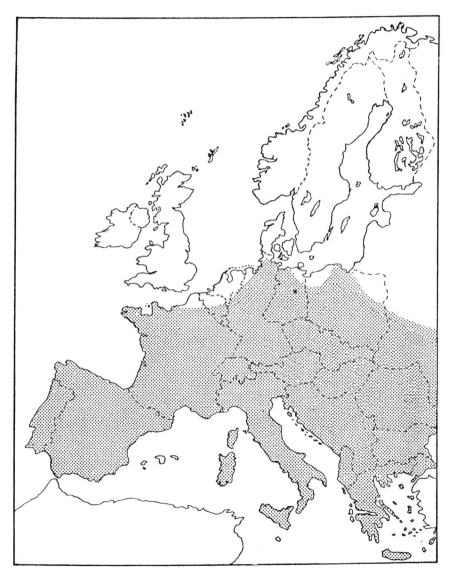

Belgium: No population data from northern areas where the species is virtually extinct (Jooris, pers. comm.). In the Grand Carrière de Romont, the last record for this species was October 1961, even though it was regularly surveyed till 1980 (94).

France: Widespread in small numbers throughout most of France.

Several very large breeding colonies numbering many hundreds of bats have been found in the Ardèche region, but these have declined due to disturbance.

In the rest of the Rhône-Alpes region, most of the observations are isolated hibernating individuals in mines and caves and one isolated record from Bretolet. A few breeding colonies are known (219).

In Ain and south Jura Mountains it is found in breeding colonies often with *Myotis brandtii*, in summer, but is usually isolated in winter. The small population has remained relatively stable for the past 20 years (Rolandez, pers. comm. 5/1981).

Netherlands: Between 1936 and 1951, over 3000 bats were ringed in the South Limburg mines and in one there was a breeding colony of 300–400 bats. This colony had used the same spot for over 35 years and it had probably been there for several hundred years. Many of the bats stayed in the same cave both in summer and winter.

Another breeding colony of about 100 bats was found in an old church near Hertogenbosch, and some of these hibernated in the South Limburg caves over 100 km to the south (24). Both these long-established colonies had died out by 1950 (49).

Other colonies disappeared in the 1950s and 1960s, and most recent records of isolated animals in hibernation come from the South Limburg caves. A few bats have occurred in church lofts in southern and south-eastern Netherlands (Glas, pers. comm.).

Germany: A population in the south numbering 4500 in 1961 had fallen 90% by 1976. The number of juveniles produced

Myotis myotis - mouse-eared bat

annually in 3 small nursery colonies was 112 in the 1950s, but by the 1960s only 13 were born (170).

In another area in southern Germany, bats hibernating in caves declined from about 300 to 50 between 1968 and 1977 (81), and nursery roosts formerly containing 800–1000 bats now have less than 250 (8).

Europe: **Czechoslovakia:** Summer colonies number 50–2000, and in winter up to 200 in the areas of highest population in central Bohemia and southern Moravia. Large declines have occurred in the last few years, especially in agricultural and industrial areas (182).

East Germany: Large populations numbering 400–500 bats in chalk tunnels around Berlin declined about 85% in 30 years (101).

Poland: There are several records of declining populations in Poland.

At Malogoszczy in the Kieleckie province, a nursery colony declined from 200 in 1962 to 25 in 1978 and only 12 in 1979 (236).

Hibernating bats in the Koralowa cave near Czestochowa declined from 100 in 1951 to 10–20 in 1966, and it is probably now extinct.

In the early 1950s there was a large nursery colony of 3000 in a Krakow church but this is now extinct (Bogdanowicz, pers. comm. 1981).

Populations in 4 large caves in the Krakow-Czestochowa Uppland have decreased substantially in the last 35 years (114).

Spain: It occurs throughout the Iberian peninsula (221).

Switzerland: A large nursery colony of several hundred bats in a building in the north-west was lost between 1945 and 1947, apparently due to building alterations (5).

World: In Israel it is on the verge of extinction (152).

THREATS

It has been collected and greatly disturbed, particularly in caves. Colonies have been killed by remedial timber treatment in buildings and others have been deliberately killed or excluded from their nursery sites, especially those in castles, churches and other large buildings. Reductions in numbers of large beetles in grassland, as well as agricultural pesticides, are probably reducing bat survival and breeding success.

CONSERVATION MEASURES

Protected in all countries. Some roost sites are specially protected in most countries. All nursery roosts should be protected both in buildings and caves – as well as the important hibernation sites.

ADDITIONAL BIBLIOGRAPHY

1, 2, 14, 17, 26, 28, 32, 33, 34, 36, 50, 57, 58, 59, 61, 62, 73, 77, 79, 80, 83, 84, 86, 87, 95, 96, 99, 103, 104, 108, 112, 115, 119, 120, 126, 129, 130, 135, 137, 138, 143, 145, 164, 166, 168, 172, 175, 181, 188, 190, 191, 192, 196, 209, 211, 216, 218, 224, 227, 235, 239.

17. *Barbastella barbastellus* - Schreber 1774

E. Barbastelle bat
F. Barbastelle d'Europe
G. Mopsfledermaus
I. Barbastello

D. Bredøret flagermus
N. Mopsvleermuis
S. Murciélago de bosque
H. Nycteris i mikromolossos

DISTRIBUTION

EEC: England and Wales, Denmark, Germany to Italy, except the south, Sicily, Corsica, Sardinia. Not found in Greece.

Denmark: Zealand, Lolland and Falster only.

Europe: Austria, Switzerland, E Germany, Czechoslovakia, Hungary, Rumania, Bulgaria, N Yugoslavia, central Spain, SE Norway and Sweden.

World: W Europe including England, Wales and S Scandinavia, to the Volga, Crimea, Caucasus, Morocco, and the large Mediterranean islands.

HABITAT

Prefers cool areas – uplands in southern Europe, lower altitudes in north. Most common in forest and riparian habitats, and often feeds low over water. Roosts in hollow trees behind loose bark and in outer areas of caves, mines, cellars and tunnels. In winter, usually solitary but some tight large clusters of several hundred are known. In summer, also forms nursery clusters in buildings.

POPULATION

EEC: Little known. In north-west Europe and Britain the species is probably in decline but there are few figures. It is very rarely found and appears to be one of Europe's rarest bats occurring widely but sparsely.

Britain: Little is known about the population, but probably occurs throughout England and Wales. Only one or 2 specimens are found each year.

Barbastella barbastellus

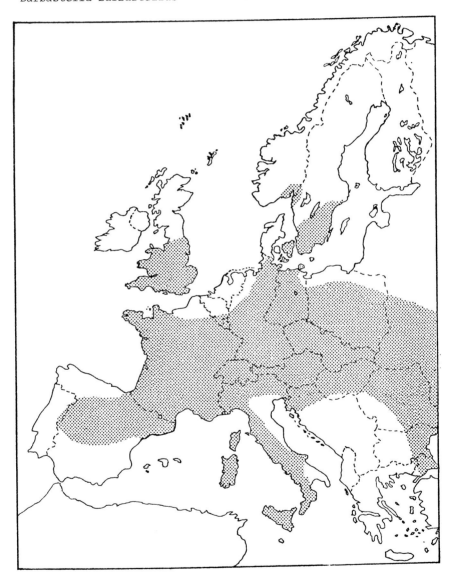

Barbastella barbastellus - barbastelle bat

Belgium: A considerable decline in population is recorded (73).

France: Very rarely recorded, nearly all are of isolated individuals. In Ain a few small winter populations are known, with a group of about 20 bats in one tunnel; but there are no summer records (Rolandez, pers. comm. 5/1981). This species is reported to be declining (Tupinier, pers. comm.).

Germany: A considerable decline in the population has occurred (Roer, pers. comm.). It is now a very rare species.

Netherlands: In the South Limburg mines only 49 bats were recorded from 1936 to 1951. Only 4 records have been made in the rest of the country (Glas, pers. comm., Lina, pers. comm., 24).

Europe: In central and east Europe it is relatively much more abundant and colonies of up to 2000 bats are found.

Czechoslovakia: Found throughout, but is more frequent in

Bohemia and Moravia than Slovakia. There have been very few summer observations, but there are records of winter populations numbering tens or hundreds. A decrease in numbers has been noted with some large declines taking place recently (182).

Hungary: Apparently very rare (Topál, pers. comm.).

Norway: Extremely rare species, only a few records from the Oslo area (Pedersen, pers. comm. 4/1981).

Spain: Few summer observations but winter records from hibernating caves are more numerous. No colonies are known (221, 43).

Sweden: A very rare species whose status is unknown (Gerell, pers. comm. 1981).

World: Unknown.

THREATS

Changes in land use and particularly loss of hollow trees are probably most important threats, together with pollution in riparian habitats.

CONSERVATION MEASURES

Protected in all countries. Research is urgently required to find status and critical habitat and to find the necessary conservation measures.

ADDITIONAL BIBLIOGRAPHY

1, 2, 5, 11, 12, 13, 14, 16, 17, 33, 34, 36, 49, 57, 61, 77, 79, 80, 81, 83, 84, 86, 103, 112, 120, 126, 129, 130, 135, 143, 150, 166, 181, 190, 191, 209, 211, 224, 232, 239.

	EEC	World
Status	V	V

18. *Plecotus auritus* - Linnaeus 1758

E. Brown long-eared bat
F. Oreillard brun
G. Braunes Langohr
I. Orecchinone

D. Langøret flagermus
N. Gewone grootoorvleermuis
S. Murciélago orejudo
H. Nycteris makrootos i europaiki

DISTRIBUTION

EEC: Throughout, except S Italy, Sicily, Corsica, Sardinia and probably Greece.

Europe: Austria, Switzerland, E Germany, Poland, Hungary, Czechoslovakia, Yugoslavia, Rumania, N Bulgaria and extreme N of Spain, S Norway, Sweden and Finland.

World: W Europe including Britain, Ireland and S Scandinavia, south to the Pyrenees, central Spain (43), central Italy, Crimea and Caucasus and east to Mongolia, SE Siberia and NW China, Sakhalin, Hokkaido and N Honshu.

HABITAT

Found in buildings, hollow trees and artificial roost boxes in summer and additionally caves, cellars and mines in winter. Lives in forests, woodland and parkland and at high altitude, especially in S Europe.

POPULATION

EEC: Forms small scattered colonies in summer containing up to 50 bats (occasionally over 100), but little is known about overall population changes. Usually solitary in hibernation. It is a rare bat in much of southern Europe.

Britain: Probably the second most abundant bat (after *P. pipistrellus*) occurring everywhere except perhaps exposed regions of NW Scotland and offshore islands (202).

Denmark: Only about 0·5% of about 4700 bats hibernating in the Smidie, Tingbaek and Mønstead mines in Jutland were *P. auritus* (Jensen, pers. comm., 12).

Plecotus auritus

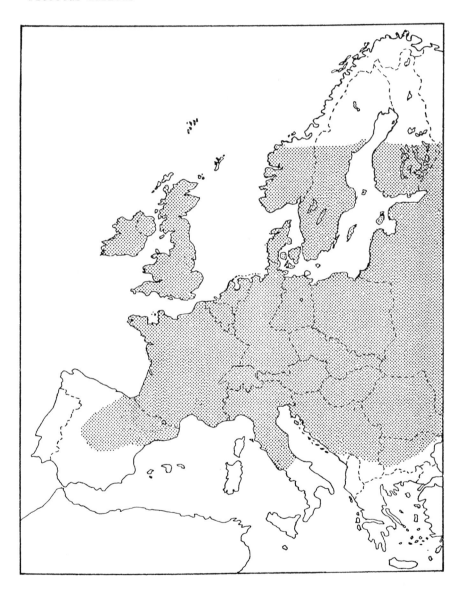

Plecotus auritus - brown long-eared bat

France: This species' distribution appears to be linked with high altitude. A small breeding colony was found at Meaudre at an altitude of 1053 metres. Small hibernating groups of 10–13 bats have been found in the forested regions of the Beaujolais Mountains between 700 and 750 metres. Generally it is not well recorded, but is thought to be fairly abundant (157, Tupinier, pers. comm., 219, 220, Rolandez, pers. comm. 5/1981).

Netherlands: Substantial declines have been recorded of hibernating bats in caves, but more significant are declines in 76 church nursery colonies. During a period of between 5 and 10 years, an overall decline of 67% was recorded (from 600 bats); 34 colonies were lost totally, with 26 greatly reduced, and the remainder (all very small colonies) remaining constant or increasing slightly (49, 37).

Europe: **Czechoslovakia:** Found throughout, except the treeless areas of southern Moravia and Slovakia. Small summer colonies are widely distributed in areas of favourable habitat (182).

Hungary: Reported to be a rare species (Topál, pers. comm.).

Norway: It is fairly abundant throughout its range and the population stable (Pedersen, pers. comm. 4/1981).

Spain: Considered abundant at the beginning of the century but since then there have been few records (221, 43, 243).

Sweden: Very abundant in southern Sweden, frequently using artificial roost boxes (Gerell, pers. comm. 1981).

World: Little known.

THREATS

Loss of hollow trees and remedial timber treatment in buildings have killed many colonies of these bats (37).

CONSERVATION MEASURES

Protected in all countries. Banning the use of toxic chemicals in buildings would be the most important conservation measure for this and most other species.

ADDITIONAL BIBLIOGRAPHY

1, 2, 5, 11, 13, 14, 17, 24, 26, 32, 33, 34, 35, 36, 38, 50, 58, 60, 65, 68, 73, 77, 78, 79, 80, 83, 84, 86, 87, 91, 96, 101, 103, 107, 109, 112, 120, 124, 126, 129, 130, 135, 137, 138, 143, 145, 147, 164, 166, 170, 172, 175, 181, 190, 191, 192, 198, 209, 211, 224, 232, 235, 239.

19. *Plecotus austriacus* - Fischer 1829

E. Grey long-eared bat
F. Oreillard gris
G. Graues Langohr

N. Grijze grootoorvleermuis
H. Nycteris makrootos i mesogiaki

DISTRIBUTION

EEC: S Britain, France, Belgium, Luxembourg, Netherlands, Germany except north, Italy, except south, Corsica and Greece. Archaeological material from northern England shows former more northerly range *c* 5000 years BP (Stebbings' data).

Europe: Austria, Switzerland, Czechoslovakia, Hungary, Rumania, Bulgaria, Yugoslavia, Albania, central and N Iberian peninsula and southern E Germany and Poland.

World: S Europe and N Africa through the Caucasus and Palestine to the Himalayas, Mongolia and W China, S England, Jersey, Canary Islands and Cape Verde Islands.

HABITAT

Mostly found in urban lowlands and highly wooded areas, especially in southern Europe. Roosts in buildings and hollow trees mainly, but also caves, cellars and mines in winter.

POPULATION

EEC: Little known. Widespread and abundant in southern areas but rare in north-west Europe where *P. auritus* is most abundant. Forms nursery colonies of up to 50 individuals.

Britain: Found only in Dorset, Hampshire, Sussex and Devon. One colony in Dorset numbered 22 in 1961, fell to 4 in the cold winter of 1962–63, but has since increased (202).

Belgium: Only record of this species was of remains found in a tawny owl pellet in 1979 in the Soignes forest (234).

France: The distribution appears to be closely linked with altitude, with it preferring lowlands.

Plecotus austriacus

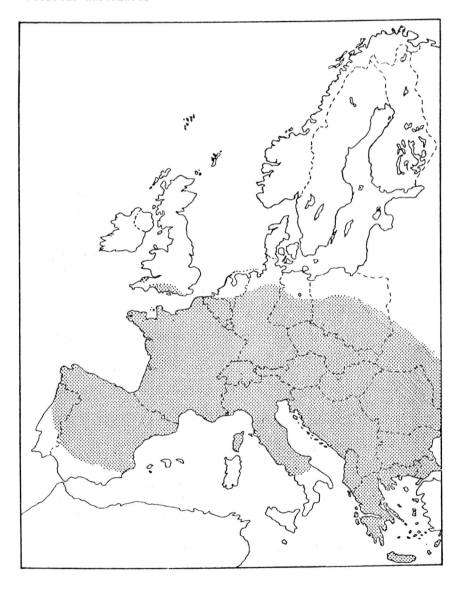

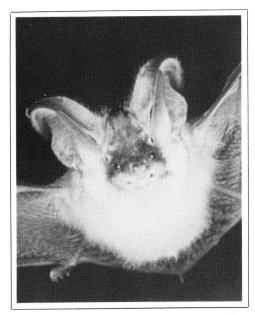

Plecotus austriacus - grey long-eared bat

Nearly all the records are of hibernating bats found in caves either on the plains or at low altitudes, eg at Dombes, 250–300 metres, at Cremieux, 230 metres.

In Ain it is regularly found in roofs in summer in small colonies, but it is most often seen in autumn (Tupinier, pers. comm., 157, 219, 220, Rolandez, pers. comm. 5/1981).

In Corsica, bats previously identified as *P. auritus* were corrected to *P. austriacus* but its status is unknown (27).

Europe: **Czechoslovakia:** Slightly more frequently found than *P. auritus*, with summer colonies numbering up to 30 bats. They hibernate singly. The winter population is decreasing (182).

World: Unknown.

THREATS

Loss of woodlands, hollow trees and remedial timber treatment in buildings pose greatest threats and causes of declines.

89

CONSERVATION MEASURES

Protected in all countries. Banning the use of toxic chemicals in remedial treatment of timber in buildings will have greatest conservation value.

ADDITIONAL BIBLIOGRAPHY

4, 5, 14, 17, 32, 33, 34, 35, 36, 38, 49, 60, 61, 68, 73, 77, 79, 80, 81, 83, 84, 86, 87, 96, 101, 103, 108, 109, 112, 115, 120, 124, 126, 129, 135, 143, 145, 166, 170, 172, 181, 190, 191, 192, 209, 211, 232, 235, 239.

20. *Miniopterus schreibersii* - Kuhl 1819

E. Bent-winged bat, Schreibers'   N. Schreiber's vleermuis
   bat   S. Murciélago de cueva
F. Minioptère de Schreibers   H. Nycteris i macropterys
G. Lanflügelfledermaus
I. Miniottero

DISTRIBUTION - partly migratory
EEC: Greece, Corsica, Sardinia, Sicily, Italy and southern France.
            Does not occur in Germany (Roer, pers. comm.).

Europe: Iberian peninsula, Switzerland, Austria, Yugoslavia, Albania,
            Rumania, Bulgaria, Hungary and southern Czechoslovakia.

            Switzerland: Recently reported from west and south only
            (Baagøe, pers. comm. 1982).

            Czechoslovakia: Southern Slovakia only (182).

World: S Europe and Morocco through Caucasus to China and Japan,
            most of Oriental region, New Guinea, Australia and Africa
            south of Sahara.

SYSTEMATICS
            Little known, but recent work worldwide suggests that a
            number of separate species will be identified eventually. In
            Europe all bats will probably conform to a single species.

HABITAT
            Roosts in subterranean sites in winter and summer, and
            occasionally occurs in buildings in summer. Mostly found in
            hilly areas and feeds over open country. Known to migrate
            200 km from winter to summer roosts (5).

POPULATION
EEC: Little known in western Europe.

            France: Substantial declines have occurred in France; one
            colony numbering 7000 became virtually extinct in 10 years in

91

Miniopterus schreibersii

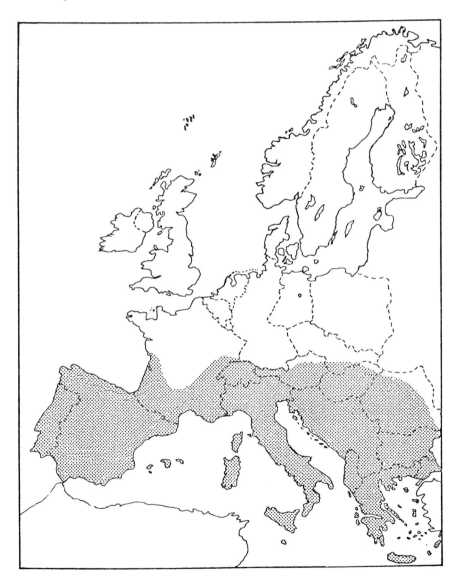

the 1950s (40). Declines in the east have also been recorded. This decline was due to frequent disturbance, but the species still breeds in Jura, and probably does in Ain even though the caves there are mostly used in winter. Currently the population appears stable (Rolandez, pers. comm. 5/1981).

Many bats migrate between states, eg France to Spain (15).

Europe: A few large colonies of several thousand individuals gathered from wide areas are known, especially in SE Europe, but little systematic work has been done and few population trends are known.

Czechoslovakia: Northern boundary of its distribution is in south Slovakia. Some caves still contain colonies numbering between 100 and 1000 bats, but generally it is a very rare, declining and threatened species (182).

Rumania: A survey of the caves in the Rumanian Dobrogea showed serious declines in numbers. In 1974 a nursery colony of 2000–3000 had declined to 100–200 bats by 1979 (44).

Spain: Some migratory movements of this bat have been discovered through ringing. It was found to migrate from the Barcelona region to the Baux de Provence on the French side of the Pyrenees, in some cases travelling through the Col de Perthus.

Virtually all the records for this species come from underground habitats (221, 43).

Switzerland: Very substantial declines in population have been recorded, eg a cave colony numbering 2000 bats was recorded from 1950 to 1960 but it became extinct in the following 10 years (5).

World: It is on the verge of extinction in Israel (152).

THREATS

Loss of caves and mines by infilling and mining and disturbance in caves by speleologists, biologists ringing bats and tourists are the most important known threats. The species is very sensitive and vulnerable to disturbance.

CONSERVATION MEASURES

The species is protected in all countries where it occurs, but individual sites are generally not specially protected. All cave roosts require protection from disturbance.

ADDITIONAL BIBLIOGRAPHY

2, 14, 17, 26, 48, 50, 59, 61, 62, 84, 86, 87, 112, 126, 143, 151, 168, 209, 211, 218, 224, 227, 239.

CHIROPTERA: VESPERTILIONIDAE   Status: EEC   World
                                                   V   V

21. *Pipistrellus pipistrellus* - Schreber 1774

E. Pipistrelle bat   D. Dvaergflagermus
F. Pipistrelle commune   N. Dwergvleermuis
G. Zwergfledermaus   S. Murciélago común
I. Pipistrello nano   H. Nycteris i kini

DISTRIBUTION

EEC: Throughout.

Europe: Throughout, including southern Scandinavia.

World: From W Europe including British Isles and S Scandinavia to the
           Volga and Caucasus, Morocco, Asia Minor and Palestine,
           Turkestan, Iran, Afghanistan and Kashmir.

HABITAT

           Found mostly in urban areas and agricultural and lightly
           wooded areas. Occurs predominantly in buildings throughout
           the year but also roosts in hollow trees and artificial roost
           boxes. Large nursery colonies occur in caves in eastern
           Europe, but it is rarely found in caves in the west.

POPULATION

EEC: Probably the most abundant bat in Europe. Colonies of many
           thousands are known but there has been a general decline
           throughout the area since 1950.

           Britain: It is thought to have declined substantially since 1960.
           Annual surveys from 1978 to 1983 of between 78 and 214
           colonies showed declines totalling 55% (206). Average colony
           size fell from about 119 to 53 bats. .

           Belgium: Populations appear stable (Fairon, pers. comm.).

           France: Abundant with breeding roosts in buildings and
           occasionally caves, eg Bourne valley (157, 219).

Europe: **Czechoslovakia:** Dependent on buildings and so is most
           common in urban areas – some towns having a density of 30

Pipistrellus pipistrellus

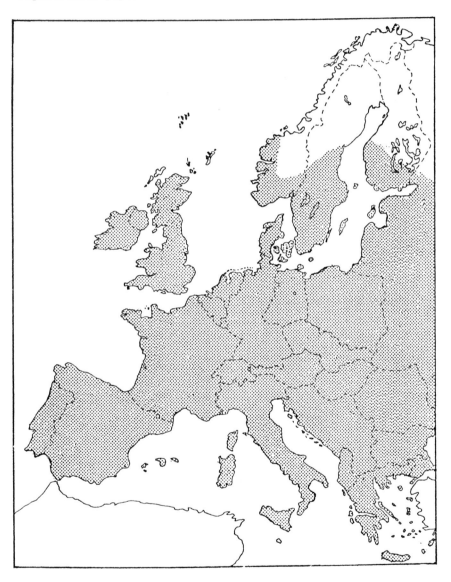

Pipistrellus pipistrellus - pipistrelle bat

km^{-2}. Summer and winter colonies usually number between 10 and 300. Sudden invasions of roosts by up to 800 bats in August and September have been recorded, linked with short migrations between summer and winter roosts. This species is rare in some areas (182, 202).

Hungary: Widely distributed, with moderate-sized populations which are reported to be stable (Topál, pers. comm.).

Norway: The population is apparently stable and may be increasing (Pedersen, pers. comm. 4/1981).

Rumania: In 1962 a large population of hibernating *Pipistrellus pipistrellus* was found in a limestone cave. Total number is estimated to be between 80 000 and 100 000. It is thought that these bats migrate to spend the summer in central Russia (64, 211).

Sweden: Very abundant in the south (Gerell, pers. comm. 1981).

World: Unknown.

THREATS

Remedial timber treatment in buildings is probably the greatest threat, together with agricultural pesticides and loss of tree roosts. Cavity wall insulation and deliberate killing are also important. Accidental deaths are often caused by bats becoming trapped in ventilation ducts and rainwater pipes (171, Stebbings' data).

CONSERVATION MEASURES

Protected in all countries. Devising new methods of remedial timber treatment of roof timbers will probably be the most effective conservation measure, as well as banning the use of existing toxic chemicals.

ADDITIONAL BIBLIOGRAPHY

1, 2, 5, 11, 12, 13, 14, 24, 26, 32, 33, 34, 36, 49, 50, 61, 62, 73, 79, 80, 86, 87, 98, 100, 103, 106, 120, 126, 128, 129, 143, 145, 164, 166, 170, 172, 187, 190, 191, 192, 198, 209, 232, 235, 239.

CHIROPTERA: VESPERTILIONIDAE   Status: R   V

22. *Pipistrellus nathusii* - Keyserling & Blasius 1839

E. Nathusius' pipistrelle bat
F. Pipistrelle de Nathusius
G. Rauhhäut Fledermaus
I. Pipistrello di Nathusius

D. Troldflagermus
N. Ruige dwergvleermuis
S. Falso murciélago común
H. Nycteris i trachydermos

DISTRIBUTION - a highly migratory species

EEC: Throughout except Ireland, Sicily, Sardinia. Two, apparently vagrants, caught in Britain.

Europe: Throughout except Scandinavia – though may be present in Sweden, Iberian peninsula – only N and central Spain, and southern Portugal.

World: From W Europe to the Urals, Caucasus and W Asia Minor.

HABITAT

Lives in parkland and mixed light woodland, often in riparian habitats. Roosts in hollow trees and buildings, summer and winter. Increasingly found in bird and bat boxes placed in woodland.

POPULATION

EEC: Little known. It is very rare in west Europe where only single specimens have been recorded. It may be more abundant than is apparent but it is known to be strongly migratory in some areas, especially in the east. There are several records of over 1000 km, the longest flight being over 1600 km (211).

Britain: Only 2 records, with some evidence that there may be more (199, 18). However, the records may be of migratory animals as both were found in the autumn.

Belgium: Two records only. A dead female bat was found in Hoevenen in September 1980, and an injured male was found a month later in a building in Ruette. The injury was caused by a ring and the bat died a few days later. It had been ringed at Teufelssee (Muggelsee area of Berlin) and had travelled a

Pipistrellus nathusii

Pipistrellus nathusii - Nathusius' pipistrelle bat

distance of around 670 km in 36 days from ringing. It seems likely that this species overwinters in Belgium regularly (75).

Denmark: Last recorded in 1957, but may still occur as a rare species (12, Baagøe, pers. comm. 1981).

France: Rarely recorded. Only one colony has been found, hibernating at Biviers in January 1976 in a hollow tree. These may have migrated from NE Europe. The only other records are of individuals found in various areas, including Corsica (157, 219, Noblet, pers. comm. 6/1983).

Netherlands: Rare, but in recent years there has been an increase in records in bird and bat boxes (Glas, pers. comm.).

Europe: **Czechoslovakia:** It is very rare, with only isolated individuals being recorded throughout the country in summer. It is thought to occur in the warmer south Slovakia lowlands throughout the year (182).

Hungary: Rarely found, with unknown status (Topál, pers. comm.).

101

Iberian peninsula: There are no recent records but this species probably occurs in small numbers in winter (221).

Sweden: Two specimens were recorded in bat boxes in August 1982 in a pine forest 20 km east of Lund (90). There are only 4 earlier records but from recent observations using a bat detector it is thought to be more widespread (Gerell, pers. comm. 1981).

World: Little known, but more plentiful in central and eastern Europe.

THREATS

Loss of hollow trees and remedial timber treatment in buildings are likely to be the most significant threats.

CONSERVATION MEASURES

Protected in all countries. Nothing is known about the ecology of this species and therefore no knowledge of what conservation measures are required.

ADDITIONAL BIBLIOGRAPHY

1, 5, 11, 13, 19, 33, 36, 49, 52, 73, 79, 80, 83, 101, 108, 117, 120, 143, 145, 164, 166, 190, 191, 192, 208, 229, 232, 239.

CHIROPTERA: VESPERTILIONIDAE   Status: V   V

23. *Pipistrellus kuhli* - Kuhl 1819

E. Kuhl's pipistrelle bat   N. Kuhl's dwergvleermuis
F. Pipistrelle de Kuhl   S. Murciélago de borde claro
G. Weiszrandfledermaus   H. Nycteris i leukogyros
I. Pipistrello albolimbato

DISTRIBUTION

EEC: Southern half of France, Italy, Sicily, Corsica, Sardinia and Greece.

Europe: Switzerland, S Austria, W Yugoslavia, Albania, and west to Istanbul, Spain, and Portugal except NW (215).

            Switzerland: Reported to occur only in the west and south (Geneva and Tessin) (Aellen, pers. comm.).

World: Southern Europe, Crimea, Caucasus and Turkestan to Pakistan throughout SW Asia and N Africa, much of Africa south of the Sahara.

HABITAT

Found in urban areas and agricultural habitats. Roosts primarily in buildings, but is also found in hollow trees.

POPULATION

EEC: Very little known.

            France: In some areas it is considered abundant, but there are very few records with only one from Isère and 2 localities in the Rhône-Alpes region – Dombes and Lyon (157, 219, Tupinier, pers. comm.).

            Italy: It has been recorded throughout except in the Basilicata region, but status is unknown (Vernier, pers. comm. 1981).

Europe: Very little known.

World: The most abundant bat in Israel where numbers remain stable (152).

Pipistrellus kuhli

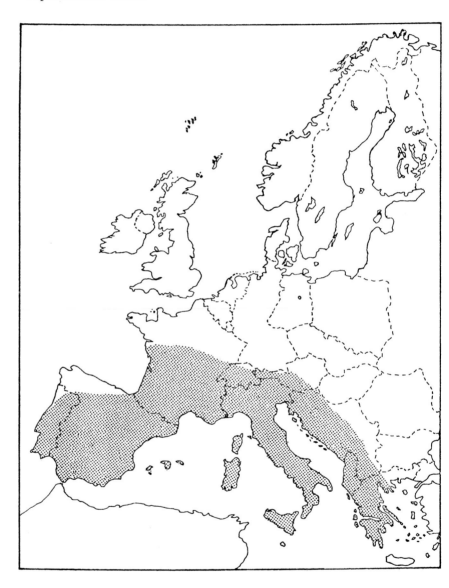

THREATS
Remedial timber treatment in buildings and loss of hollow trees will be the major threats.

CONSERVATION MEASURES
Protected in all countries where it occurs. We have no knowledge of its ecology and requirements. Banning the use of toxic chemicals in buildings will be the most effective conservation measure.

ADDITIONAL BIBLIOGRAPHY
50, 62, 78, 143, 145, 187, 239.

24. *Pipistrellus savii* - Bonaparte 1837

E. Savi's pipistrelle bat
F. Pipistrelle de Savi
G. Alpenfledermaus
I. Pipistrello di Savi

N. Savi's dwergvleermuis
S. Murciélago montãnero
H. Nycteris i alpios

DISTRIBUTION

EEC: SE France, Italy, Corsica, Sardinia, Sicily, S Germany and Greece.

Europe: Spain, Portugal, Switzerland, Austria, W Yugoslavia, Albania, SW Bulgaria.

Switzerland: Found in west (Geneva), the Alps, and south (Tessin) (Aellen, pers. comm.).

World: From Iberia, Morocco, the Canary and Cape Verde Islands through the Crimea, Caucasus, Asia Minor, Turkestan and Mongolia to Korea and Japan, south-eastwards through Iran and Afghanistan to Punjab.

HABITAT

Mostly found in mountainous areas near woodland and buildings. Roosts mainly in buildings and rock crevices but also in hollow trees. Hibernates in outer parts of caves and in buildings.

POPULATION

EEC: Very little known.

France: Generally very rare in France (Tupinier, pers. comm.). First recorded in Isère in 1978 (Noblet, pers. comm. 6/1983).

Corsica: Probably fairly abundant but few records (23).

Italy: It has been recorded throughout except for the SE region (Basilicata) (Vernier, pers. comm. 1981).

Pipistrellus savii

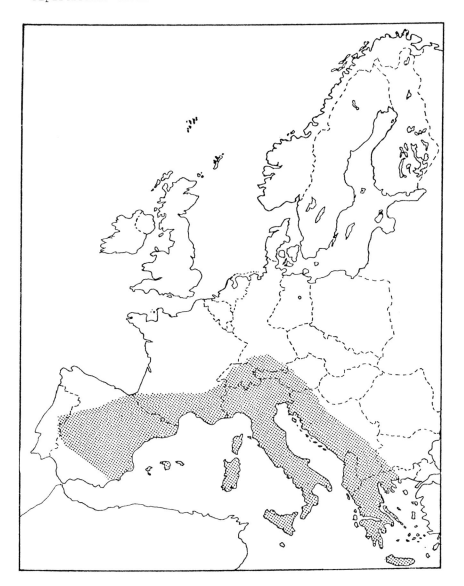

Europe: Very few records.

World: Nothing known.

THREATS
Remedial timber treatment in buildings poses greatest threat.

CONSERVATION MEASURES
Protected in all countries where it occurs.

ADDITIONAL BIBLIOGRAPHY
62, 145, 158, 239.

25. *Eptesicus serotinus* - Schreber 1774

E. Serotine bat
F. Grande sérotine
G. Breitflügelfledermaus
I. Serotino comune

D. Sydflagermus
N. Laatvlieger
S. Murciélago hortelano
H. Nycteris i eurypteryx

DISTRIBUTION
EEC: From southern Britain and throughout region.

Britain: Since 1970 it has been found further north and west, but this may represent an increase in observers rather than a true extension of range.

Denmark: Appears to have increased its range northwards (13).

Europe: Throughout except Norway and Finland but only recently found in Sweden, near Kristianstad (89).

World: From W Europe through southern Asiatic Russia to the Himalayas, Thailand and China, north to Korea, also in N Africa and most of the Mediterranean islands.

HABITAT
Winter roosts are in hollow trees and buildings, and in some areas, especially those with very cold winters, it also occurs in caves and mines. In summer, colonies are mostly found in urban buildings but it also occurs in hollow trees. In the Netherlands it is thought to be found only in buildings (Glas, pers. comm.). It feeds predominantly on large insects in open sheltered urban and parkland areas, mostly in lowlands.

POPULATION
EEC: Most nursery colonies number fewer than 100 bats but some colonies are known containing several hundred bats. Population sizes in buildings appear to be stable in much of Europe.

Britain: Some nursery colonies are known to have declined

Eptesicus serotinus

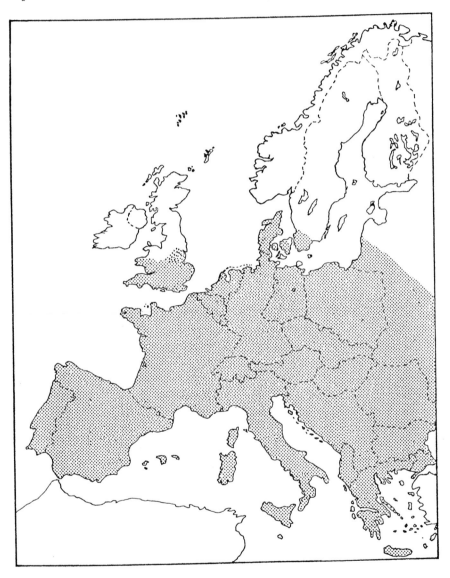

substantially since 1960 (from 100+ to about 25 bats) (unpublished data).

Denmark: Dispersed throughout except north Zealand, with colonies usually numbering up to 50 bats (Baagøe, pers. comm. 1981, 13). During the last 100 years the Serotine has colonized Jutland, Funnen, Lolland-Falster, and has only been recorded in Zealand since 1973.

France: Rarely found, with only a few isolated records (157, Rolandez, pers. comm. 5/1981, 219, 239).

Germany: Large declines in numbers have been recorded for nursery roosts in buildings in NW Germany. Five nursery colonies showed an average reduction from 45 to 5 bats between the mid-1950s and 1960s (172). In northern North-rhine-Westfalia it remains abundant (Taake, pers. comm.).

Netherlands: A study of 21 colonies in the 1960s and 1970s showed little overall change. Some colonies increased while others decreased or disappeared (96). Nearly all the colonies number less than 100 bats (Glas, pers. comm.).

Europe: European populations appear to be stable.

Czechoslovakia: Found throughout except in mountainous regions, but most abundant in central Moravia (182, 202).

Hungary: Widely distributed, but generally small populations. Although regarded as non-migratory, a movement of 144 km was recorded (Topál, pers. comm., 202).

Spain: There are very few records, but it is thought to be widespread and abundant.

Sweden: First recorded in October 1982 near Kristianstad in the southernmost part of Sweden. There were about 10 identified using a bat detector (89).

World: Little known.

111

THREATS

The species is heavily dependent on buildings and remedial timber treatment is the greatest threat and known cause of death (96). Changes in land use resulting in the reduction of numbers of large beetles (which constitute the major food) may be causing population declines.

CONSERVATION MEASURES

Protected in all countries. Nothing known about critical habitat and conservation requirements, but banning the use of toxic chemicals in remedial timber treatments in buildings will be the most effective conservation measure.

ADDITIONAL BIBLIOGRAPHY

2, 5, 11, 12, 14, 24, 31, 32, 33, 34, 36, 49, 62, 73, 79, 80, 81, 83, 84, 86, 87, 101, 103, 108, 122, 126, 129, 130, 135, 143, 145, 158, 164, 166, 170, 181, 190, 191, 192, 209, 211, 232, 235, 236.

26. *Eptesicus nilssonii* - Keyserling & Blasius 1839

E. Northern serotine bat
F. Sérotine de Nilsson
G. Nordfledermaus
I. Serotino di Nilsson

D. Nordflagermus
N. Noordse vleermuis
S. Murciélago norteño
H. Nycteris i borios

DISTRIBUTION - partly migratory

EEC: South Germany, extreme eastern France and northern Italy.
            Vagrants may occur anywhere.

Europe: Scandinavia, Switzerland, Austria, Czechoslovakia, E Germany,
            Poland, E Hungary and north Rumania.

World: Central and E Europe to E Siberia, north to beyond the Arctic
            Circle in Scandinavia, south to Iraq, the Elburz Mountains,
            Pamirs and Tibet.

HABITAT

Lives in upland regions in central Europe and at lower altitude
further north. An arboreal species roosting mostly in trees but
also occurring in buildings and occasionally rock crevices and
caves.

POPULATION

EEC: A rarely found bat in west and central Europe, with no large
            colonies known. It probably has a low population density but it
            may be more abundant than is apparent.

            France: No records except from Col de Bretolet. It is thought
            more likely to be found in Jura than Alsace (219, Tupinier, pers.
            comm.).

            Italy: Only one record near Tret in 1929 (Vernier, pers. comm.
            1981).

Europe: **Czechoslovakia:** Generally rare but found throughout in
            mountainous regions. Nursery colonies of up to 40 individuals
            are known with similar-sized hibernating groups (182).

113

Eptesicus nilssonii

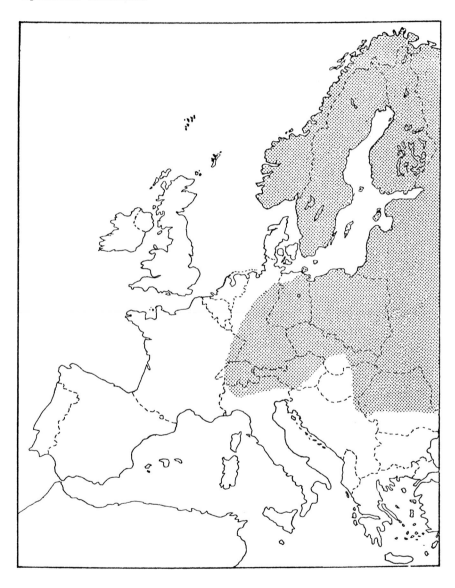

Hungary: Very rare. It is thought there are no nursery colonies (Topál, pers. comm.).

Scandinavia: The most abundant species with populations stable or increasing (Gerell, pers. comm. 1981, Pedersen, pers. comm. 4/1981).

World: Unknown.

THREATS

Loss of hollow trees and remedial timber preservation in buildings pose greatest threats.

CONSERVATION MEASURES

The species is protected in all areas where it occurs. Little is known of its status and habitat requirements. Some hibernation roosts are protected in Czechoslovakia and Poland (182, Bogdanowicz, pers. comm. 1981).

ADDITIONAL BIBLIOGRAPHY

1, 2, 11, 12, 13, 84, 87, 126, 129, 137, 138, 143, 146, 156, 161, 181, 192, 211, 216, 228, 230, 236, 239.

27. *Vespertilio murinus* - Linnaeus 1758

E. Parti-coloured bat
F. Sérotine bicolore
G. Zweifarbfledermaus
I.   Serotino bicolore

D. Skimmelflagermus
N. Tweekleurige vleermuis
S. Murciélago bicolor
H. Nycteris i dichromos

DISTRIBUTION - a strongly migratory species, especially in eastern Europe
EEC: In Denmark (NW Zealand), S Germany, E France, N Italy and
               Greece. Vagrants may occur in all EEC countries.

Europe: Austria, Switzerland, Hungary, Czechoslovakia, Yugoslavia,
               Albania, Rumania, Bulgaria, Poland, E Germany and southern
               Scandinavia.

World: From central Europe through S Siberia to the Ussuri region of
               Siberia, S Scandinavia, south to Iran and Afghanistan.

HABITAT
               Little known. Roosts in buildings, hollow trees and rock
               crevices in towns and in mountains.

POPULATION
EEC: Generally little known, with only single specimens being found
               in southern and western areas. Small colonies are found
               predominantly in upland and northern areas. It is a very mobile
               species which may turn up almost anywhere. Few bats are
               likely to breed in western Europe where they hibernate.
               Nursery roosts are mostly in north-eastern Europe.

               Denmark: Common in N Zealand, especially in buildings.
               Isolated records from Jutland (12).

               France: There are a few records from Jura and Isère. Ringed
               bats recovered up to 130 km away suggest movements along
               the plain of Gresivaudan (157, Noblet, pers. comm. 10/1982).
               The bats may breed in northern Europe.

Europe: A widespread species in eastern Europe which migrates long
               distances between summer and winter roosts (211).

Vespertilio murinus

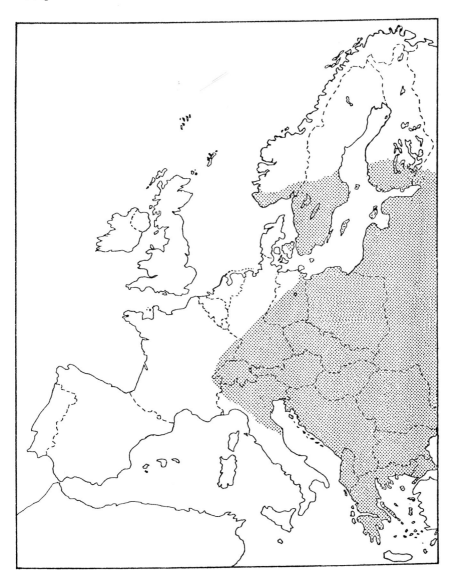

Czechoslovakia: Isolated records show this to be a very rare species (182).

Hungary: A rare species which probably has no nursery colonies (Topál, pers. comm.).

Norway: The population appears stable (Pedersen, pers. comm. 4/1981).

Sweden: Abundant in southern parts but status little known (Gerell, pers. comm. 1981).

World: Unknown.

THREATS
Loss of roosts and remedial timber treatment in buildings are likely threats.

CONSERVATION MEASURES
Protected in all countries. Nothing is known of its food and habitat requirements.

ADDITIONAL BIBLIOGRAPHY
1, 5, 11, 13, 49, 52, 53, 87, 101, 129, 143, 145, 158, 164, 192, 209, 215, 236, 239.

28. *Nyctalus leisleri* - Kuhl 1818

E. Leisler's bat
F. Noctule de Leisler
G. Kleiner Abendsegler

N. Bosvleermuis
S. Murciélago nocturno
H. Nycteris pterygisti i mikra

DISTRIBUTION - a migratory species, especially in central and eastern Europe

EEC: Ireland, Wales, southern half of Britain, S Germany, and central France, Italy and Greece. Vagrants occur widely. There are no records for Luxembourg but almost certainly occurs there.

Belgium: Only one record from Villiers-la-Ville but the identification is doubtful (Jooris, pers. comm.).

Netherlands: First recorded in 1981 near Nijmegen (Voûte, pers. comm. 3/1982).

Europe: Austria, Switzerland, Czechoslovakia, Hungary, Albania, W Yugoslavia, Bulgaria, Rumania, southern E Germany and Poland.

World: From W Europe to the Urals and Caucasus, Britain, Ireland, Madeira and the Azores, the western Himalayas and E Afghanistan.

HABITAT

Found in deciduous and conifer forests as well as parkland and urban areas. Roosts in hollow trees mainly, and artificial roost boxes, as well as buildings. It often occurs in new as well as old buildings.

POPULATION

EEC: Little known. Usually forms small colonies in trees, sometimes up to 100 bats, but when it occurs in buildings colonies of several hundred individuals are known. It appears to be very rare throughout much of the region.

Ireland: One of the most abundant species – perhaps the stronghold in its world distribution. A nursery colony, in a

Nyctalus leisleri

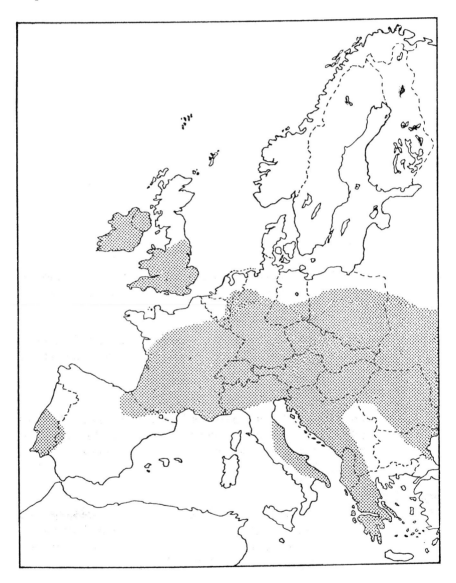

Nyctalus leisleri - Leisler's bat

house, of about 500 in south-west Eire declined to about 400 from 1969 to 1981 (Stebbings' data).

Britain: Recorded in a few buildings with colonies up to 40 bats. Otherwise very rare, only averaging 2 or 3 records each year, mostly from bat boxes in conifer plantations.

France: There are very few records and it may even be limited to the northern Alps regions. Recorded in Isère, in the Rhône-Alpes region at the Col de Bretolet and in Ain (157, Rolandez, pers. comm. 5/1981, 219, Tupinier, pers. comm.).

Netherlands: First recorded in 1981 when a colony of adult females and juveniles was found near Nijmegen in eastern Netherlands (Voûte, pers. comm. 3/1982, Glas, pers. comm.).

Europe: Little known. Appears to be a rare species throughout eastern Europe.

Czechoslovakia: A very rare bat with few records. The largest colony found contained 17 bats (182).

121

Hungary: Very rare (Topál, pers. comm.).

Spain: A rare species. It is known in central Spain, and has been recorded at Caceres, and Sierra de Gredos (Noblet, pers. comm. 6/1981, 221, Tupinier, pers. comm.).

World:
Azores: A conspicuous species as it tends to fly diurnally, probably due to lack of raptors (155). Population thought to be stable.

Status elsewhere unknown.

THREATS

Colonies in buildings are threatened by remedial timber treatment, and other colonies by loss of forest, hollow trees and parkland. Reduction in numbers of large insects may also be affecting populations.

CONSERVATION MEASURES

Protected in all countries. Artificial roost boxes readily adopted, especially in conifer forests. Banning the use of toxic chemicals in remedial timber treatment in buildings will probably be an important conservation measure.

ADDITIONAL BIBLIOGRAPHY

1, 73, 79, 80, 86, 87, 108, 116, 129, 143, 145, 158, 192, 209, 216, 236, 239.

CHIROPTERA: VESPERTILIONIDAE   Status: V(?E in V

29. *Nyctalus noctula* - Schreber 1774

E. Noctule bat   D. Brunflagermus
F. Noctule   N. Rosse vleermuis
G. Abendsegler   S. Murciélago nocturno
I. Nottola   H. Nycteris pterygisti i esperobios

DISTRIBUTION - a migratory species, especially in central and eastern
Europe

EEC: Throughout, except Ireland and northern Scotland.

Europe: Throughout, but only southernmost part of Scandinavia.

World: Europe, including most of Britain, S Scandinavia and most of
the Mediterranean islands, to the Urals and Caucasus, Morocco, SE Asia Minor to Palestine, W Turkestan to the Himalayas
and China, Taiwan, N Honshu, Japan.

HABITAT

An arboreal species preferring deciduous woodland and
parkland but it is also found in conifer forests. Roosts mainly in
hollow trees but also occurs in artificial roost boxes, in
buildings and exceptionally in rock clefts and caves. Feeds on
large beetles, moths and crickets, caught both aerially and
picked off the ground.

POPULATION

EEC: Little known. Typically forms summer colonies of 30–100 bats
in hollow trees, but much larger colonies in buildings. In NW
Europe, eg Netherlands and Britain, it is now becoming rare
with substantial declines since 1950.

Britain: No recent records from Scotland except vagrants in
extreme north (202). Observations suggest substantial and
rapid declines but there is no adequate documentation.

France: Status is unknown. Observations are few with only 2
records in Isère – both from caves, post-1978, and in Ain and at

Nyctalus noctula

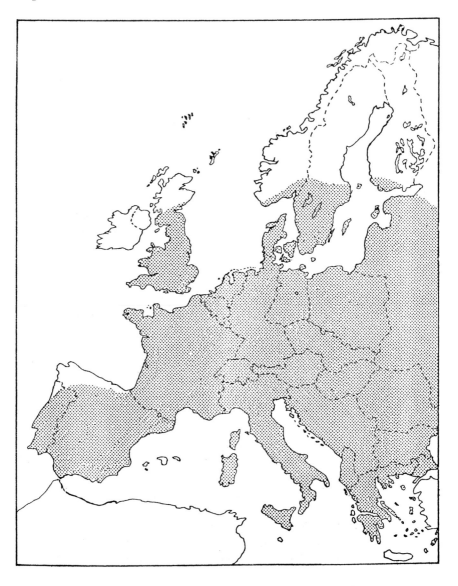

Nyctalus noctula - noctule bat

Col de Bretolet and Saleve (157, Noblet, pers. comm. 4/1981, Rolandez, pers. comm. 5/1981, 219, Tupinier, pers. comm.). It is likely to be much more abundant than records suggest.

Germany: Large nursery roosts in Bavaria numbering 800–1000 bats declined to less than 160 bats within 15 years (8).

Europe: Hibernating colonies in excess of a 1000 bats are known in several eastern European countries.

Czechoslovakia: An abundant species with stable populations found throughout, except in mountainous regions. Nursery colonies number 10–100 bats, hibernating clusters 50–1000 (182).

Hungary: The population is small, but widely distributed and apparently stable (Topál, pers. comm.).

Spain: Rarely recorded but probably occurs throughout in small groups (221).

Sweden: Abundant in the south, but is regarded as vulnerable because of the rapid destruction of hollow tree roosts (Gerell, pers. comm. 1981).

World: Unknown

THREATS
Many are killed during the felling of hollow trees and branches, and the reduction in number of large beetles probably contributes to declines.

CONSERVATION MEASURES
Protected in all countries. Management of agricultural habitats, such as permanent pasture, to encourage the increase in the numbers of large beetles, eg *Melolontha* spp., may be the key to this bat's conservation. Protection of hollow trees and provision of artificial roosts (bat boxes) are also required.

ADDITIONAL BIBLIOGRAPHY
1, 2, 5, 11, 12, 13, 21, 24, 26, 33, 34, 36, 49, 57, 73, 79, 80, 83, 86, 87, 101, 105, 108, 118, 120, 129, 143, 145, 158, 164, 166, 168, 190, 191, 192, 209, 232, 235, 239.

CHIROPTERA: VESPERTILIONIDAE   Status:

	EEC	World
Status:	R	R

30. *Nyctalus lasiopterus* - Schreber 1780

E. Greater noctule
F. Grand noctule
G. Riesenabendsegler

I. Nottola gigante
H. Nycteris pterygisti i megali

DISTRIBUTION - migrant

EEC: N Italy, Sicily, Greece and France (very few records).

Europe: Spain, Switzerland, western Austria and Yugoslavia, Albania, eastern Czechoslovakia, Hungary, Rumania, Bulgaria, Poland.

Switzerland: Found only in the Alps and Tessin (Aellen, pers. comm.).

Czechoslovakia: Single records from Slovakia, and Tatra (Aellen, pers. comm., 182).

Poland: One record from a barn owl pellet in Krolikow in the Konin district (Bogdanowicz, pers. comm. 1981, Bogdanowicz & Ruprecht, pers. comm. 10/1982, 174).

Spain: One record from Linares de Riofrio (Salamanca) (221, Tupinier).

World: From western Europe to the Urals, Caucasus, Asia Minor, Iran and the Ust-Urt Plateau.

HABITAT

Not known. Probably roosts in trees.

POPULATION

EEC: It is a very rare bat known only from a few scattered records.

France: Very rare. Only known from a few old records in the south, and at the Col de Cour on the French-Swiss border (Haute-Savoie) (157, 219, Tupinier, pers. comm.).

Nyctalus lasiopterus

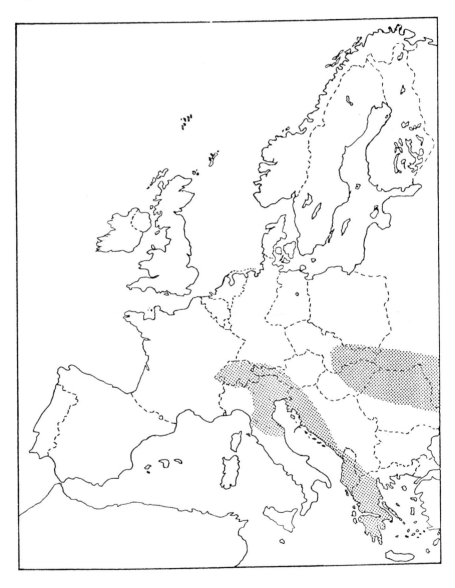

However, Aellen (pers. comm.) suggests that there are many more records for France.

Italy: It is thought to breed in Italy but there are no positive records.

Europe: **Bulgaria:** Very rare with a few scattered records (Aellen, pers. comm.).

Hungary: Only 3 records, of which one was a pregnant female which gave birth while in captivity (215, Topál, pers. comm.).

World: A very rare bat throughout its range, about which almost nothing is known.

THREATS

Loss of roosts and reduction in numbers of large insects will probably be greatest threats.

CONSERVATION MEASURES

Protected in countries where it has occurred. A very rare bat about which nothing is known. It could not be studied in EEC countries because it is too rare. No colonies are known in Europe.

ADDITIONAL BIBLIOGRAPHY

None.

31. *Tadarida teniotis* - Rafinesque 1814

E. European free-tailed bat
F. Molosse de Cestoni
G. Bulldoggfledermaus
I. Molosso di Cestoni

N. Bulvleermuis
S. Murciélago rabudo
H. Nycteris i urophoros

DISTRIBUTION - a migratory species

EEC: Sicily, Corsica, Sardinia, Italy, SE France, Greece and Crete.

Europe: Iberian peninsula except north-west and Pyrenees, Switzer-land, W Austria, NW and SE Yugoslavia and southern Bulgaria.

World: Mediterranean Europe and most Mediterranean islands and Madeira, Morocco and Algeria, Egypt and Asia Minor east to Kirghizia and Afghanistan, also eastern Asia from eastern Himalayas through China to N Korea and Japan (3).

HABITAT

Feeds on aerial insects often at great height. In summer and winter, roosts in buildings, rock clefts and occasionally large high caves and mines.

POPULATION

EEC: Little known because it has not been studied. Said to form large colonies in Europe, which is typical for the genus elsewhere.

France: This Mediterranean species has only been recorded in the south-east. In Isère in 1975, 18 bodies were found in a cave in the Bourne valley; no juvenile skeletons were found so it is not thought to have been a breeding colony (163).

There are 3 records from the Rhône-Alpes region. At Vallon at the Pont d'Arc several were seen flying, one was shot at Villebois (south of Bugey) and a few were netted at the Col de Bretolet (219). No nursery colonies are known (Rolandez, pers. comm. 5/1981).

Occurs in Corsica but status is unknown (23).

Tadarida teniotis

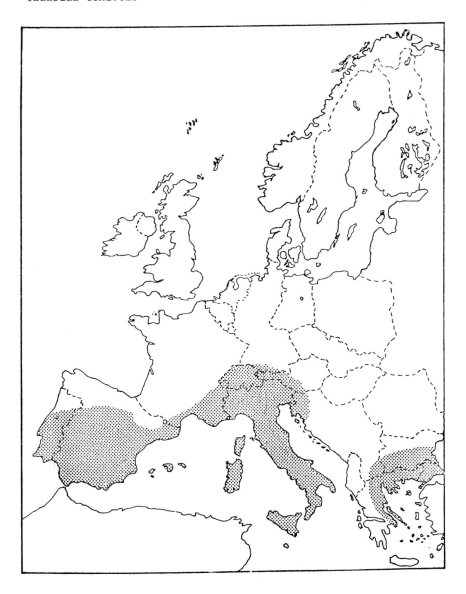

Europe: **Iberian peninsula:** Early this century it was abundant throughout, except in the north-west. It is now considered rare (221).

World: Unknown.

THREATS
Remedial timber treatment in buildings, disturbance of cave
colonies and reductions in numbers of large insects are
probably the main threats, but little is known.

CONSERVATION MEASURES
Protected in all countries where it occurs. Nothing is known
about status and critical habitat. Research is urgently required.
All nursery roosts in caves need protection from disturbance.

ADDITIONAL BIBLIOGRAPHY
1, 52, 61, 62, 145, 158.

BIBLIOGRAPHY

1. **Aellen, V.** 1961. Le baguement des chauves-souris au Col de Bretolet (Valais). *Archs Sci., Genève,* **14,** 365-392.

2. **Aellen, V.** 1965. Les chauves-souris cavernicoles de la Suisse. *Int. J. Speleol.,* **1,** 269–278.

3. **Aellen, V.** 1966. Notes sur *Tadarida teniotis* (Raf.) (Mammalia, Chiroptera) – 1. Systematique paleontologie et peuplement, repartition geographique. *Revue suisse Zool.,* **73,** 119–159.

4. **Aellen, V.** 1971. La chauve-souris *Plecotus austriacus* (Fischer) en Suisse. *Proc. Congr. Suisse Speleologie, 4th, Neuchatel, 1970,* 167-172.

5. **Aellen, V.** 1978. Les chauves-souris du Canton de neuchatel, Suisse (Mammalia, Chiroptera). *Bull. Soc. neuchâtel. Sci. nat.,* **101,** 5-26.

6. **Ahlen, I.** 1979. Dammfladdermus *Myotis dasycneme* (Boie 1825) funnen i Uppland. *Fauna Flora, Upps.,* **74,** 259–262.

7. **Almaca, C.** 1967. Nouvelles observations sur *Rhinolophus mehelyi,* Matschie 1901 au Portugal. *Arq. Mus. Bocage,* **1,** 35–38.

8. **Antoni, W.** 1979. Zur Gefährdung der Fledermäuse in Bayern. *Jahrb. Ver. Schutz Bergwelt,* **44,** 171–189.

9. **Atallah, S. I.** 1977 (1978). Mammals of the eastern Mediterranean region: their ecology, systematics and zoogeographical relationships. *Saugetierkd. Mitt.,* **26,** 1–50.

10. **Baagøe, H. J.** 1973. Taxonomy of two sibling pieces of bats in Scandinavia *Myotis mystacinus* and *Myotis brandti* (Chiroptera). *Vidensk. Meddr dansk naturh. Foren.,* **136,** 191–216.

11. **Baagøe, H. J.** 1980. Status for danske flagermus. In: *Status over den denske plante-og dyreverden.* Fredningsstyrelsen, 360–368.

12. **Baagøe, H. J.** 1981. Danish bats, status and protection. *Myotis,* **18–19,** 16–18.

13. **Baagøe, H. J. & Jensen, B.** 1973. The spread and present occurrence of the Serotine *(Eptesicus serotinus)* in Denmark. *Period biol.,* **75,** 107–109.

14. **Balcells, R. E.** 1961. Murciélagos del Norte Central Espanol. *Bol. Sancho Sabio,* **5,** 1–30.

15. **Balcells, R. E.** 1962. Migration en Espagne des Minioptères francais. *Spelunca, Paris,* **2,** 92–98.

16. **Balcells, R. E.** 1965. Nuevos datos sobre murciélagos raros en cuevas espanolas. *Miscelánea zool.,* **2,** 149–160.

17. **Balcells, R. E.** 1968. Nuevas citas de murciélagos y nicteribidos del pais vasco-cantabrico. *Boln R. Soc. esp. Hist. nat.,* **66,** 17–38.

18. **Banks, C., Clark, M. & Newton, R.** 1983. A second Nathusius' *(Pipistrellus nathusii)* in Britain, caught in flight in Hertfordshire. *Trans. Herts. nat. Hist. Soc.,* **29,** 15–18.

19. **Barbu, P.** 1968. O colonie estivala de *Pipistrellus nathusii* Keys., et Blas. 1839 in farul de la Sf. Gheorgie - Dobrogea. *Ocrot. Nat.,* **12,** 211–215.

20. **Barbu, P.** 1974. Ocrotirea Liliecilor. *Ocrot. Nat.,* **18,** 29–36.

21. **Barbu, P. & Sin, G.** 1968. Observatii asupra hibernarii speciei *Nyctalus noctula* (Schreber 1774) in Faleza Lacului Razelm - capul Dolosman - Dobrogea. *Studii Cerc. Biol. Ser. Zool.,* **20,** 291–297.

22. **Beaucornu, J. C.** 1965. Captures de *Myotis blythi oxygnathus* (Monticelli, 1885) (Chir. Vesp.) en Anjou et en Touraine confirmation de sa presence en Corse. *Mammalia,* **29,** 54–60.

23. **Beaucornu, J. C., Launay, H. & Noblet, J. F.** 1983. Nouvelles données sur les Chiroptères de Corse. *Mammalia,* **47,** 125–127.

24. **Bels, L.** 1952. Fifteen years of bat banding in the Netherlands. *Publtiës natuurh. Genoot. Limburg,* **5,** 1–99.

25. **Beron, P.** 1961. Contribution à la connaissance des chauves-souris Bulgares. *Fragm. balcan.*, **3**, 189–195.

26. **Beron, P.** 1963. Le baguage des chauves-souris en Bulgarie de 1940 à 1961. *Acta theriol.*, **7**, (4), 33–49.

27. **Beskov, V. & Beron, P.** 1962. Notizen über die Verbreitung und die Biologie einiger seltener Fledermäuse in Bulgarien. *Izv. zool. Inst., Sof.*, **12**, 35–39.

28. **Bezem, J. J., Sluiter, J. W. & Heerdt, P. F. van.** 1960. Population statistics of five species of the bat genus *Myotis* and one of the genus *Rhinolophus*, hibernating in the caves of S. Limburg. *Archs néerl. Zool.*, **13**, 511–539.

29. **Bezem, J. J., Sluiter, J. W. & Heerdt, P. F. van.** 1964. Some characteristics of the hibernating locations of various species of bats in South Limburg. *Proc. Akad. wet. Amst., C*, **67**, 235–350.

30. **Bogdanowicz, W. & Urbanczyk, Z.** 1981. A record of *M. nathalinae*, Tupinier 1977 from Poland. *Acta theriol.*, **26**, 427.

31. **Braaksma, S.** 1968. Nieuwe gegevens over de verspreiding van de Laatvlieger, *Eptesicus serotinus* (Schreb.), in Nederland. *Levende Nat.*, **71**, 181–188.

32. **Braaksma, S.** 1969. Summer resorts of bats on lofts and towers of churches in the Netherlands. *Lynx, Prague*, **10**, 7–12.

33. **Braaksma, S.** 1970. The distribution of bats in the Netherlands. *Bijdr. Dierk.*, **40**, 10–12.

34. **Braaksma, S.** 1973. Some details about the occurrences of bats in summer and winter resorts in the Netherlands and about the risk caused by wood-preservation activities in buildings. *Period. biol.*, **75**, 125–128.

35. **Braaksma, S.** 1980. Further details on the distribution and protection of bats in the Netherlands. *Proc. int. Bat Res. Conf., 5th, 1980*, 179–183.

36. **Braaksma, S. & Glas, G. H.** 1974. Gegevens over de achteruitgang van grootoorvleermuizen (Genus *Plecotus*) op zomerverblijfplaatsen in Nederland. *Lutra*, **16**, 24–33.

37. **Braaksma, S. & Glas, G. H.** 1980. Overwinterende vleermuizen in enkele ijs-en kasleelkelders. *Lutra*, **22**, 78–84.

38. **Bree, P. J. H. van & Dulić, B.** 1963. Notes on some specimens of the genus *Plecotus*, Geoffroy 1818 (Mammalia, Chiroptera) from the Netherlands. *Beaufortia*, **10**, 7–18.

39. **Brink, F. H. van den.** 1967. *A field guide to the mammals of Britain and Europe*. London: Collins.

40. **Brosset, A.** 1966. *La biologie des Chiroptères*. Masson.

41. **Brosset, A.** 1974. *Mammifères sauvages de France*. Natham.

42. **Brosset, A.** 1978. Les chauve-souris disparaissent-elles? *Courr. Nature*, no. 55, 17–22.

43. **Carbonell, M.** 1979. Anillamientos y controles de quiropteros obtenidos en la Boveda, en la Granja de san Ildefonso, Segovia. *Bol. Estac. cent. Ecol.*, **16**, 67–72.

44. **Cerveny, J.** 1982. Notes on the bat fauna (Chiroptera) of Roumanian Dobrogea. *Nyctalus*, n.s. **1**, (4/5), 349–357.

45. **Cerveny, J. & Horacek, I.** 1980-81. Comments on the life history of *Myotis nattereri* in Czechoslovakia. *Myotis*, **18–19**, 156–162.

46. **Clark, D. R.** 1981. Bats and environmental contaminants: a review. *Spec. scient. Rep. U.S. Fish Wildl. Serv., Wildl.*, no. 235.

47. **Corbet, G. B.** 1978. *The mammals of the Palaearctic region: a taxonomic review*. London: British Museum; Ithaca: Cornell University Press.

48. **Crucitti, P.** 1976. Biometrics of a collection of *Miniopterus schriebersi* (Chiroptera) captured in Latiom, Italy. *Annali Mus. civ. Stor. nat. Giacomo Doria*, **81**, 131–138.

49. **Daan, S.** 1980. Long-term changes in bat populations in the Netherlands: a summary. *Lutra*, **22**, 95–105.

50. **Deblase, A. F. & Martin, R. L.** 1973. Distributional notes on bats (Chiroptera Rhinolophidae, Vespertilionidae) from Turkey. *Mammalia*, **37**, 598–602.

51. **Degn, H. J.** 1981. Vandflagermusens *(Myotis daubentoni)* forekomst i et vestfynsk omrade. *Flora Fauna,* **81,** 3–6.

52. **Desmet, J. F. & Noblet, J. F.** 1976. Données sur la Pipistrelle de Nathusius, *Pipistrellus nathusii* (Keyserling et Blasius), le Molosse de Cestoni, *Tadarida teniotis* (Rafinesque), et la Sérotine Bicolore, *Vespertilio murinus* (Linne) dans le departement de L'Isère. *Mammalia,* **40,** 521–523.

53. **Dick, D.** 1982. Zum Vorkommen der Zweifarbfledermaus (*Vespertilio discolor* Kuhl) im Bezirk Karl-Marx-Stadt. *Nyctalus,* n.s. **1,** (4/5), 447–448.

54. **Dinale, G.** 1965. Alcuni risultati dell 'inanellamento di *Rhinolophus ferrumequinum* Schreber e di *Rhinolophus euryale* Blasius in Liguria (1957–64) e nel Lazio (1962–1965). *Boll. Zool.,* **32,** 815–822.

55. **Dinale, G.** 1966. Studi sui Chirotteri Italiani: V. Esperimenti di ritorno al luogo di cattura e ricatture esterne di *Rhinolophus ferrumequinum* Schreber inanellati in Liguria. *Atti Soc. ital. Sci. nat.,* **105,** 147–157.

56. **Dinale, G.** 1967. Studi sui Chiroterri Italiani: VIII. Spostamenti di *Rhinolophus euryale* Blasius inanellati in Liguria. *Atti Soc. ital. Sci. nat.,* **106,** 275–282.

57. **Dinale, G. & Ghidini, G. M.** 1966. Centro inanellamento *Pipistrellus:* otto anni di attivita (1957–1964). *Atti Soc. ital. Sci. nat.,* **105,** 91–101.

58. **Dorgelo, J. & Punt, A.** 1969. Abundance and internal migration of hibernating bats in an aritificial limestone cave (Sibbergroeve). *Lynx, Prague,* **10,** 101–125.

59. **Dulic, B.** 1963. Étude écologique des chauves-souris cavernicoles de la Croatie Occidentale (Yougoslavie). *Mammalia,* **27,** 385-436.

60. **Dulić, B.** 1980. Morphological characteristics and distribution of *Plecotus auritus* and *Plecotus austriacus* in some regions of Yugoslavia. *Proc. int. Bat Res. Conf., 5th, 1980,* 151–161.

61. **Dulić, B. & Mikuska, J.** 1966. Two new species of bats (Mammalia, Chiroptera) from Macedonia with notes on some other bats occurring in this territory. *Fragm. balcan.,* **6,** 1–13.

62. **Dulić, B. & Tvrtkovic, N.** 1970. The distribution of bats on the Adriatic islands. *Bijdr. Dierk,* **40,** 17–20.

63. **Dumitrescu, M., Orghidan, T. & Tanasachi, J.** 1958. Pestera de la Gura Dobrogei. (The cave Pestera de la Gura Dobrogei). *Anu. Com. Geol., Bucaresti,* **31,** 461–499.

64. **Dumitrescu, M. Tanasachi, J. & Orghidan, T.** 1963. Raspindirea chiroptereol in R. R. Romina. (Distribution of bats in Romania). *Lucr. Inst. Speol. 'Emil Racovita' Acad.,* **1–2,** 509–575.

65. **Egsbaek, W., Kresten, K. & Roer, H.** 1971. Beringungsergebnisse an der Wasserfleder-maus *(Myotis daubentoni)* und Teichfledermaus *(Myotis dasycneme)* in Jutland. *Decheniana,* **18,** 51–55.

66. **Eleder, P.** 1977. A find of *Rhinolophus ferrumequinum* (Schreber 1774) in the Ceskomoravska vrchovina Highland. *Vertebr. Zpr.,* 63–71.

67. **Faber, T. & Meisch, C.** 1978. Bilan provisoire de recensement des populations de chauves-souris au Grand Duche de Luxembourg. *Bull. Inst. r. Sci. nat. Belg.,* **5,** 68–95.

68. **Fairon, J.** 1975. Presence de *Plecotus austriacus* en Belgique. *Bull. Inst. r. Sci. nat. Belg.,* **51,** 1–6.

69. **Fairon, J.** 1977. Le petit Rhinolophe *(Rhinolophus hipposideros)* (Bechstein, 1800), Chiroptère en voie de disparition? *Naturalistes belg.,* **58,** 212–225.

70. **Fairon,** 1978. Revision des Pipistrelles de la fauna belge, conservées dans les collections de L'IRSNB. *Bull. Inst. r. Sci. nat. Belg.,* **5,** 52-67.

71. **Fairon, J.** 1979. Les chauves-souris, les hommes, et l'environnement. *Oiseau,* (April-June), 81–85.

72. **Fairon, J.** 1980. *Myotis brandti* en Belgique. *Bull. Inst. r. Sci. nat. Belg.,* **52,** 1-8.

73. **Fairon, J., Gilson, R., Jooris, R., Faber, T. & Meisch, C.** 1982. Cartographie provisoire de la faune Chiroptèrologique Belgo-Luxembourgeoise. *Bull. Inst. r. Sci. nat. Belg.*, **7**, 1–125.

74. **Fairon, J., Gilson, R., Jooris, R. & Lebrun, M.** 1978. Les reserves Cheiroptèrologiques en Belgique. *Bull. Inst. r. Sci. nat. Belg.*, **5**, 13–51.

75. **Fairon, J. & Jooris, R.** 1980. *Pipistrellus nathusii* en Belgique. *Bull. Inst. r. Sci. nat. Belg.*, **6**, 40–41.

76. **Feldmann, R.** 1967. Bestandsentwicklung und heutiges Areal der Klein-hufeisennase, *Rhinolophus hipposideros* (Bechstein, 1800), im mittleren Europa. *Saugetierkd. Mitt.*, **15**, 43–49.

77. **Felten, H.** 1971. Fledermaus-Beringung im weiteren Rhein-Main-Gebiet 1959/60–1969/70. *Decheniana*, **18**, 83–93.

78. **Felten, H. & Storch, G.** 1970. Kleinsauger von den Italienischen mittelmeer-inseln pantelleria und lampedusa (Mammalia). *Senckenberg. biol.*, **51**, 159–173.

79. **Fischer, J. A.** 1982. Zum vorkommen der Fledermäuse im Bezirk Suhl - Teil I. *Nyctalus*, n.s. **1**, (4/5), 361–379.

80. **Fischer, J. A.** 1982. Zum vorkommen der Fledermäuse im Bezirk Suhl - Teil 2. *Nyctalus*, n.s. **1**, (4/5), 411–424.

81. **Frank, M., Nagel, A. & Weigold, H.** 1980. Bestandsentwicklung der in Höhlen überwinternden Fledermäuse auf der Schwabischen Alb. *Höhle*, **31**, 111–116.

82. **Gaisler, J.** 1971. Zur Okologie von *Myotis emarginatus* in Mitteleuropa. *Decheniana*, **18**, 17–82.

83. **Gaisler, J.** 1973. Netting as a possible approach to study bat activity. *Period. biol.*, **75**, 129–134.

84. **Gaisler, J.** 1975. A quantitive study of some populations of bats in Czechoslovakia (Mammalia, Chiroptera). *Acta Sci. nat. Acad. Sci. bohemoslov. Brno*, **9**, 1-44.

85. **Gaisler, J.** 1979. Ecology of bats. In: *Ecology of small mammals*, edited by D. M. Stoddart, 281–342. London: Chapman & Hall.

86. **Gaisler, J. & Hanák, V.** 1969. Ergebnisse der Zwanzigjahrigen Beringung von Fledermäusen (Chiroptera) in der Tschechoslowakei: 1948–1967. *Acta Sci. nat. Acad. Sci. bohemslov. Brno*, **3**, 1–33.

87. **Gaisler, J. & Hanák, V.** 1969. Summary of the results of bat banding in Czechoslovakia, 1948–1967. *Lynx, Prague*, n.s. **10**, 25–34.

88. **Gauckler, A. & Kraus, M.** 1970. Kennzeichen und Verbreitung von *Myotis brandti*. *Z. Saugetierk*, **35**, 113–124.

89. **Gerell, R., Ivarsson, A. & Lundberg, K.** 1983. Sydfladdermus, *Eptesicus serotinus* Schreber 1774, ny fladdermusart i Sverige. *Fauna Flora, Upps.*, **78**, 38–40.

90. **Gerell, R. & Lundberg, K.** 1983. Troll fladdermus, *Pipistrellus nathusii* Keyserling & Blasius, after antraffad i Sverige. *Fauna Flora, Upps.*, **78**, 35–38.

91. **Gilson, R.** 1973. Observations sur les Chiroptères de la Carrière Roosburg. *Bull. Soc. r. Belge Etud. Geol. Archaeol.*, **22**, 351–366.

92. **Gilson, R.** 1976. Relations entre les conditions climatiques du milieu souterrain et l'ecologie de *Myotis mystacinus* Kuhl. *Bull. Soc. r. Belge Etud. Geol. Archaeol.*, **23**, 265–280.

93. **Gilson, R.** 1980. Observations sur les Chiroptères de la Grande Carrière de Romont à Eben-Emael. *Bull. Inst. r. Sci. nat. Belg.*, **6**, 25–38.

94. **Gilson, R.** 1980. Rapports d'activitées. *Bull. Inst. r. Sci. nat. Belg.*, **6**, 66–73.

95. **Gilson, R. & Hubart, J. M.** 1973. Protection du biotype de la Grotte Lyell. *Bull. Soc. r. Belge Etud. Geol. Archaeol.*, **22**, 367–381.

96. **Glas, G. M. & Braaksma, S.** 1980. Aantalsontwikkelingen in Zomerverblijf-plaatsen van Vleermuizen in Kerken. *Lutra*, **22**, 84–95.

97. **Greece.** 1981. *Presidential decree no. 67: Protection of flora and fauna and regulations on scientific research*, 214–223.

98. Grimmberger, E. 1979. Untersuchungen über den Einfluss klimatischer Faktoren auf das Verhalten der Zwergfledermaus, *Pipistrellus pipistrellus* (Schreber 1774), im Winterquatier und während der sogenannten Invasionen. *Nyctalus*, n.s. **1,** (2), 145–157.

99. Grimmberger, E. 1980. Nördlichster Fundort vom Mausohr, *Myotis myotis* (Borkhausen 1797), und Wochenstube der Grossen Bartfledermaus, *Myotis brandti* (Eversmann 1845) in Mecklenburg. *Nyctalus,* n.s. **1,** (3), 190–192.

100. Grimmberger, E. & Bork, H. 1978. Untersuchungen zur Biologie, Okologie und Populationsdynamik der Zwergfledermaus *Pipistrellus p. pipistrellus* (Schreber 1774) in einer grossen Population im Norden der DDR. *Nyctalus,* n.s. **1,** (2), 122–136.

101. Haensel, J. 1972. Weitere Notizen über im Berliner Stadtgebiet aufgefundene Fledermäuse (Zeitraum 1967–1971). *Milu,* **3,** 303–327.

102. Haensel, J. 1972. Zum Vorkommen der beiden Bartfledermausarten in den kalkstollen von Rüdersdorf (Vorläufige Mitteilung). *Nyctalus,* no. 4, 5–7.

103. Haensel, J. 1973. Ergebnisse der Fledernausberingungen im Norden der DDR, unter besonderren Berüksichtigung des Massenwinterquartiers Rüdersdorf. *Period. biol.,* **75,** 135–143.

104. Haensel, J. 1978. Searching for intermediate quarters during seasonal migrations in the large mouse-eared bat *(Myotis myotis)*. *Proc. int. Bat Res. Conf., 4th, 1975,* 231–237.

105. Haensel, J. 1979. Abendsegler *(Nyctalus noctula)* überwintert in einem Keller. *Nyctalus,* n.s. **1,** (2), 137–138.

106. Haensel, J. 1979. Ergänzende Fakten zu den Wanderungen in Rudersdorf überwinternder Zwergfledermäuse *(Pipistrellus pipistrellus)*. *Nyctalus,* n.s. **1,** (2), 85–90.

107. Haensel, J. 1979. Invasionsartiger Einflug von Braunen Langohren, *Plecotus auritus,* in ein Gebaude der Stadt Nauen. *Nyctalus,* n.s. **1,** (2), 95–96.

108. Haensel, J. 1982. Weitere Notizen über im Berliner Stadtgebiet auf-gefundene Fledermaus (Zeitraum 1972–1979). *Nyctalus,* n.s. **1,** (4/5), 425–444.

109. Hanák, V. 1969. Okologische Bemerkungen zur Verbreitung der Langohren (Gattung *Plecotus* Geoffroy, 1818) in der Tschechoslowakei. *Lynx, Prague,* n.s. **10,** 35–39.

110. Hanák, V. 1970. Notes on the distribution and systematics of *Myotis mystacinus* Kuhl, 1819. *Bijdr. Dierk.,* **40,** 40–44.

111. Hanák, V. 1971. *Myotis brandti* (Eversmann, 1845) (Vespertilionidae, Chiroptera) in der Tschechoslowakei. *Věst. čsl. Spol. zool.,* **35,** 175–185.

112. Hanák, V. & Gaisler, J. 1970. Comments on the protection of bats in Czechoslovakia and some suggestions on the research on bat populations. *Bijdr. Dierk.,* **40,** 5–7.

113. Harmata, W. 1968. Kilka uwag na temat ochrony nietoperzy w Polsce. *Chrońmy Przyr. ojcz.,* **24,** (2), 32–36.

114. Harmata, W. 1981. Zmiany w liczebnosci Nietoperzy (Chiroptera) w niektorych jaskiniach jury Krakowsko-Czestochowskiej w Latach 1945–1979. *Rocz. Muz. Czest,* **11,** 24–30.

115. Harrison, D. L. 1951. The bats of the Segeberg-Gipshohle in Schleswig-Holstein. *Bonn. zool. Beitr.,* **1–2,** 9–16.

116. Heise, G. 1982. Nachweis des Kleinabendseglers *(Nyctalus leisleri)* im Kreis Prenzlaw, Uckermark. *Nyctalus,* n.s. **1,** (4/5), 449–452.

117. Heise, G. 1982. Zu Vorkommen, Biologie und Okologie de Ranhhautfledermaus *(Pipistrellus nathusii)* in der Umgebung von Prenzlau (Uckermark) Bezirk Neubrandenburg. *Nyctalus,* n.s. **1,** (4/5), 281–300.

118. Heise, G. & Schmidt, A. 1979. Wo uberwintern im Norden der DDR beheimatete Abendsegler *(Nyctalus noctula)?* *Nyctalus,* n.s. **1,** (2), 81–84.

119. Henkel, F. & Tress, C. & H. 1982. Zum Bestandsruckgang der Mausohren *(Myotis myotis)* in Südthürngen. *Nyctalus,* n.s. **1,** (4/5), 453–471.

120. Hiebsch, H. 1971. Bericht über die Fledermausmarkierung in den Jahren 1969–1971. *Nyctalus,* no. 3, 55–59.

137

121. **Hildenhagen, U. & Taake, K. H.** 1981. Westfalens grösste derzeit bekannte Fledermaus-Winterquartiere an der Westfalischen Pforte. *Nat. Heim.*, **41**, 59–62.

122. **Hildenhagen, U. & Taake, K. H.** 1982. Zur Bestandssituation und Biologie der Breitflügelfledermaus *Eptesicus serotinus* (Schreber 1774) im nordöstlichen Westfalen. *Nat. Heim.*, **42**, 21–26.

123. **Hohne, M.** 1981. Zwischenbericht zum Fledermausschutz - program Nordbaden. *Veröff. Naturschutz Landschaftspflege Baden-Württ.*, **53/54**, 245–273.

124. **Horáček, I.** 1975. Notes on the ecology of bats of the genus *Plecotus* Geoffroy 1818 (Mammalia: Chiroptera), *Věst. čsl. Spol. zool.*, **39**, 195–210.

125. **Horáček, I., Ceveny, J., Tausl, A. & Vitek, D.** 1974. Notes on the mammal fauna of Bulgaria (Insectivora, Chiroptera, Rodentia). *Věst. čsl. Spol. zool.*, **38**, 19–31.

126. **Horáček, I. & Zima, J.** 1978. Net-revealed cave visitation and cave-dwelling in european bats. *Folia zool. Praha*, **27**, 135–148.

127. **Horáček, I. & Zima, J.** 1979. Zur Frage der Synanthropie bei Hufeisennasen in der Tschechoslowakei. *Nyctalus*, n.s. **1**, (2), 139–141.

128. **Hůrka, L.** 1966. Beitrag zur Bionomie, Okologie und zur Biometrik der Zwergfledermaus *(Pipistrellus pipistrellus* Schreber 1774) nach den Beobachtungen in Westbohmen. *Věst. čsl. Spol. zool.*, **30**, 228–246.

129. **Hůrka, L.** 1983. Die Bewertung des vorkommens der Fledermäuse (Mammalia: Chiroptera) in Westböhmen. *Věst čsl. Spol. zool.*, **47**, 31–35.

130. **Husson, A. M.** 1954. A preliminary note on the bats hibernating in the casemates of the town of Luxembourg. *Arch. Inst. Luxemb. Sci. nat. phys. math.*, n.s. **21**, 65–70.

131. **Huttere, R.** 1978. Ein weiter Nachweis de Kleinen Wasserfledermaus *Myotis nathalinae. Bonn. zool. Beitr.*, **29**, 1–3.

132. **Jefferies, D. J.** 1972. Organochlorine insecticide residues in British bats and their significance. *J. Zool.*, **166**, 245–263.

133. **Kahmann, H. & Caglar, M.** 1960. Turkiyede mameli hayvanlar arastirimi sahasinda yori buluslar. *Biyol. Derg.*, **10**, 119–126.

134. **Kahmann, H. & Goerner, P.** 1956. Les chiroptères de Corse. *Mammalia*, **20**, 333–389.

135. **Karlstedt, K.** 1972. Zur Fledermausfauna der Heimkehle bei Uftrungen. *Nyctalus*, no. 4, 8–10.

136. **Kirk, G.** 1971. Gesetzlicher Fledermausschutz in Europa. *Decheniana*, **18**, 45–50.

137. **Knolle, F.** 1977. Zum vorkommen, zum Überwinterungsverhalten sowie zur Bestandsent-wicklung der Fledermäuse im niedersachsischen Harz. *Beitr. Nat. kd. Niedersachs.*, **3**, 49–57.

138. **Knolle, F.** 1978. Über Massnahmen zur Erhaltung und Sicherung von Fledermauswinter-quartieren in Harz. *Jahrb. Ver. Schutz Bergwelt*, **43**, 193–196.

139. **Kowalski, K.** 1953. Material relating to the distribution and ecology of cave bats in Poland. *Fragm. faun.*, **6**, 541–567.

140. **Kowalski, K.** 1953. Nietoperze jas kiniowe Polski i ich ochrona. (Cave dwelling bats in Poland and their protection). *Ochr. Przyr.*, **21**, 59–77.

141. **Kowalski, K.** 1955. Our bats and their protection. *Zakl. Ochr. Przyr. Krakow*, **11**, 1–110.

142. **Krzanowski, A.** 1961. Wyniki rozwieszenia skrzynek dla nietoperzy w Bialowieskim Paku Narodowym. *Chronmy Przyr. ojcz.*, **17**, 29–32.

143. **Krzanowski, A.** 1973. Numerical comparison of Vespertilionidae and Rhinolophidae (Chiroptera: Mammalia) in the owl pellets. *Acta zool. cracov.*, **18**, 133–140.

144. **Krzanowski, A.** 1980. *Nietoperze.* Warszawa.

145. **Laar, V. van & Daan, S.** 1964. On some Chiroptera from Greece. *Beaufortia*, **10**, 158–166.

146. **Lanza, B.** 1957. Su alcuni chirotteri della penisola balcanica presenza di *Eptesicus nilssoni nilssoni* (Keyserling e Blasius, 1839) in Iugoslavia. *Monitore zool. ital.*, **65**, 3–6.

147. **Lanza, B.** 1960. Su due specie criptiche di Orecchione: *Plecotus auritus* (L.) e *P. wardi* Thomas (Mamm. Chiroptera). *Monitore zool. ital.*, **68**, 7–23.

148. **Libois, R. M. & Vranken, M.** 1981. *Myotis bechsteini* en Corse. *Mammalia*, **45**, 380–381.
149. **Lina, P. H. C.** 1980. De Wettelijke Bescherming van Vleermuizen in Europe. *Lutra*, **22**, 5–7.
150. **Lina, P. H. C.** 1980. Zomervondst van een Mopsvleermuis *(Barbastella barbastellus* Schreber 1774) te S-Gravenhage. *Lutra*, **23**, 1–2.
151. **Loriol, B. de.** 1960. Observations sur les populations de *Miniopterus schreibersi* Kuhl (Chiroptère) de l'est de France. *Proc. Congr. Soc. Savantes, 84th, 1959*, 673–677.
152. **Makin, D.** 1979. The quest for Israel's bats. *Israel Land Nat.*, **5**, 28–34.
153. **Masing, M.** 1980. Tommulendlane Eestis. *Esti Loodus*, 29–34.
154. **Miller, G. S.** 1912. *Catalogue of the mammals of western Europe*. London: British Museum (Natural History).
155. **Moore, N. W.** 1975. The diurnal flight of the Azorean bat *(Nyctalus azoreum)* and the avifauna of the Azores. *J. Zool.*, **177**, 483–506.
156. **Mosansky, A. & Gaisler, J.** 1965. Ein Beitrag zur Erforschung der Chiropterenfauna der Hohen Tatra. *Bonn. zool. Beitr.*, **3/4**, 249–267.
157. **Noblet, J. F.** 1978. Les chauves-souris du departement de l'Isère. *Bull. Soc. dauphin. Etud. biol.*, **1978**, 71–82.
158. **Noblet, J. F.** 1980. Nouvelles données sur la repartition des Chiroptères de la region Rhône-Alpes. Cas particulier du departement de l'Isère. *Ciconia*, **4**, 5–11.
159. **Noblet, J. F. & Barnet D.** 1980. Les Chiroptères de la Vallée du fango (Corse) 1976–1979. *Bull. Soc. Sci. Hist. Nat. Corse.*, no. 637, 99–102.
160. **Nyholm, E. S.** 1965. Zur Okologie von *Myotis mystacinus* (Leisl.) und *Myotis daubentoni* (Leisl.) (Chiroptera). *Ann. zool. fenn.*, **2**, 77–123.
161. **Ohlendorf, B.** 1980. Zur Verbreitung der Nordfledermaus, *Eptesicus nilssoni* (Keyserling u. Blasius 1839) im Harz nebst Bemerkungen über Schutz, Uberwinterungsverhalten und Vergleiche zu anderen Fledermausarten. *Nyctalus*, n.s. **1**, (3), 253–262.
162. **Ohlendorf, B. & G., Hackethal, H., Schroder, J., Fischer, J. A. & Weber B.** 1982. Kleine Mitteilungen. *Nyctalus*, n.s. **1**, (4/5), 472–477.
163. **Palmeirim, J. M.** 1978. First records of *Myotis blythi* Tomes 1857 (Chiroptera) from Portugal. Its systematics and distribution in the Iberian Peninsula. *Arq. Mus. Bocage*, **6**, 311–318.
164. **Pieper, H. & Wilden, W.** 1980. Die Verbreitung der Fledermäuse (Mamm.: Chiroptera) in Schleswig-Holstein und Hamburg 1945–1979. *Faunistisch-Okol. Mitt.*, Suppl. 2, 3–31.
165. **Punt, A.** 1970. Round table discussion on bat conservation - summary. *Bijdr. Dierk.*, **40**, 3.
166. **Punt, A., Bree, P. J. H. van, De Vlas, J. & Wiersema, G. J.** 1974. De Nederlandse Vleermuizen. *Wet. Meded. K. ned. natuurh. Veren.*, no. 104, 1–48.
167. **Randik, A.** 1969. Ochrana netopierov na Slovensku. (Bat protection in Slovakia). *Lynx, Prague*, **10**, 79–84.
168. **Roer, H.** 1971. Weitere Ergebnisse und Aufgaben der Fledermausberingung in Europa. *Decheniana*, **18**, 121–144.
169. **Roer, H.** 1975. Zur Verbreitung und Okologie der grossen Bartfledermaus *Myotis brandti* (Eversmann 1845) im mitteleuropäischen Raum. *Säugetierkd. Mitt.*, **23**, 138–143.
170. **Roer, H.** 1977. Zur Populationsentwicklung der Fledermäuse (Mammalia: Chiroptera) in der Bundesrepublik Deutschland unter besonderer Berucksichtigung der Situation im Rhein-land. *Z. Säugetierk.*, **42**, 265–278.
171. **Roer, H.** 1979. 1180 Zwergfledermäuse *(Pipistrellus pipistrellus* Schreber) in Entluftungs-srohren eines Gebäudes verenet. *Myotis*, **17**, 31–40.
172. **Roer, H.** 1980. Population trends of bats in the Federal Republic of Germany with particular reference to the Rhineland. *Proc. int. Bat Res. Conf., 5th, 1980*, 193–197.
173. **Romero, D. & Castroviejo, J.** 1973. El quiroptero *Rhinolophus blasii*, nuevo mamifero para la fauna Iberica. *Boln R. Soc. esp. Hist. nat.*, **71**, 309–310.

174. **Ruprecht, A. L.** 1970. Borowiec olbrzymi, *Nyctalus lasiopterus* (Schreber, 1780) nowy ssak w faunie Polski. *Acta theriol.*, **15**, 370–372.

175. **Ruprecht, A. L.** 1971. Distribution of *Myotis myotis* (Borkhausen, 1797) and representatives of the genus *Plecotus* Geoffroy, 1818 in Poland. *Acta theriol.*, **16**, 95–104.

176. **Ruprecht, A. L.** 1974. The occurrence of *Myotis brandti* (Eversmann, 1845) in Poland. *Acta theriol.*, **19**, (6), 81–90.

177. **Ruprecht, A. L.** 1979. Bats (Chiroptera) as constituents of the food of barn owls *(Tyto alba)* in Poland. *Ibis*, **121**, 489–494.

178. **Ruprecht, A. L.** 1981. Variability of Daubentons bats and distribution of the Nathalinae morphotype in Poland. *Acta theriol.*, **26**, (22), 349–357.

179. **Russel, F. & Wilhelm, M.** 1972. Die gross Bartfledermaus (*Myotis brandti* Eversmann 1845) im Osterzgebirge gefunden. *Nyctalus*, **3**, 64.

180. **Rybář, P.** 1973. Remarks on banding and protection of bats. *Period. biol.*, **75**, 177-179.

181. **Rybář, P.** 1975. Hibernation of the Barbastelle, *Barbastella barbastellus* (Schreber 1774) in a man-made hibernational quarter. *Zool. listy*, **24**, 113–124.

182. **Rybář, P.** 1981. *A draft of the red data list of bats in Czechoslovakia*. Pardubice: Regional Centre for the Protection of Historical Monuments and Nature Conservation. (Unpublished.)

183. **Ryberg, O.** 1947. *Studies on bats and bat parasites*. Stockholm: Svensk Naturv.

184. **Ryberg, O.** 1947. Skanes fladdermoss. *Skanes Nat.*, 289–299.

185. **Rzebik-Kowalska, B., Woloszyn, B. W. & Nadachowski, A.** 1978. A new bat, *Myotis nattereri* (Kuhl 1818) (Vespertilionidae) in the fauna of Iraq. *Acta theriol.*, **23**, (37), 541–545.

186. **Saint-Girons, M.-C.** 1973. *Les Mammifères de France et du Benelux (faune marine exceptée)*. Paris: Doin.

187. **Saint-Girons, M.-C.** 1981. Les Pipistrelles et la circulation routière. *Mammalia*, **45**, (1).

188. **Schierer, A., Mast, J.-Cl. & Hess, R.** 1972. Contribution a l'étude eco-ethologique du Grand Murin (*Myotis myotis*). *Terre Vie*, **26**, 38–53.

189. **Schmidt, A.** 1979. Sommernachweise der grossen Bartfledermaus *(Myotis brandti)* im Kreis Beeskow Bezirk Frankfurt/0. *Nyctalus*, n.s. **1**, (1), 158–160.

190. **Schmidt, A.** 1980. Zum Vorkommen der Fledermäuse im Suden des Bezirkes Frankfurt/0. *Nyctalus*, n.s. **1**, (3), 209–226.

191. **Schober, W.** 1976. Fledermausvorkommen im Bezirk Leipzig. *Nyctalus*, no. 5, 19–24.

192. **Schober, W., Haensel, J., Handtke, K., Natuschke, G., Knorre, D. V., Stratmann, B., Wilhelm, M. & Zimmermann, W.** 1971. Zur Verbreitung der Fledermäuse in der DDR (1945-1970). *Nyctalus*, no. 3, 1–50.

193. **Siivonen, L.** 1979. *Phjolan nisakkaat. Mammals of Northern Europe.* Moscow.

194. **Skuratowicz, W.** 1948. Uwagi o ochronie niektorych ssakow. *Chrońmy Przyr. ojcz.*, **3–4**, 8–18.

195. **Sluiter, J. W. & Heerdt, P. F. van.** 1958. Observations écologiques sur quelques colonies estivales de chauves-souris, des grottes en France. *Notes biospéol.*, **13**, 111–120.

196. **Sluiter, J. W. & Heerdt, P. F. van.** 1964. Distribution and abundance of bats in S. Limburg from 1958 till 1962. *Natuurh. Maandbl.*, **53**, 164–173.

197. **Sluiter, J. W., Heerdt, P. F. van & Voûte, A. M.** 1971. Contribution to the population biology of the pond bat, *Myotis dascyneme* (Boie, 1825). *Decheniana*, **18**, 1–44.

198. **Sponselee, G. M. P., Glas, G. H. & Wiersema, G. J.** 1973. Einige gegevens over Vleermuizensterfte bij bespuiting van Kerkzolders. *Lutra*, **15**, (1-3), 1–5.

199. **Stebbings, R. E.** 1970. A bat new to Britain *Pipistrellus nathusii* with notes on its identification and distribution in Europe. *J. Zool.*, **161**, 282–286.

200. **Stebbings, R. E.** 1971. Bat protection and the establishment of a new cave reserve in the Netherlands. *Stud. Speleol.*, **2**, 103–108.

201. **Stebbings, R. E.** 1971. Bats in danger. *Oryx*, **10**, 311–312.

202. **Stebbings, R. E.** 1977. Order Chiroptera. In: *The handbook of British mammals*, edited by G. B. Corbet & H. N. Southern, 2nd ed., 68–128, 471–477. Oxford: Blackwell Scientific.

203. **Stebbings, R. E.** 1978. An outline global strategy for the conservation of bats. *Proc. int. Bat Res. Conf., 5th, 1980*, 173–178.

204. **Stebbings, R. E.** 1980. Bats. *World Wildl. Yearb.*, 1978–1979, 181–183.

205. **Stebbings, R. E.** 1982. Radio tracking greater horseshoe bats with preliminary observations on flight patterns. In: *Telemetric studies of vertebrates*, edited by C. K. Cheeseman & R. B. Mitson. *Symp. zool. Soc. Lond.*, **49**, 161–173.

206. **Stebbings, R. E. & Arnold, H. R.** 1982. Bats - an insecticide under threat? *Nat. Devon*, **3**, 7–26.

207. **Stebbings, R. E. & Jefferies, D. J.** 1982. *Focus on bats: their conservation and the law.* London: Nature Conservancy Council.

208. **Straeten, E. van der, Jooris, R. & Stuyck, J.** 1981. Eerste vondst van nathusius' dwergvleermuis *Pipistrellus nathusii* in Belgie. *Lutra*, **24**, (1), 1–6.

209. **Stratmann, B.** 1979. Untersuchungen über die historische und gegenwärtige Verbreitung der Fledermäuse im Bezirk Halle (Saale) nebst Angaben zur Okologie. *Nyctalus*, n.s. **1**, (2), 97–121.

210. **Stratmann, B.** 1980. Untersuchungen über die historische und gegenwärtige Verbreitung der Fledermäuse im Bezirk Halle (Saale) nebst Angaben zur Okologie. *Nyctalus*, n.s. **1**, (3), 177–186.

211. **Strelkov, P. P.** 1969. Migratory and stationary bats (Chiroptera) of the European part of the Soviet Union, *Acta zool. cracov.*, **14**, 393–440.

212. **Strelkov, P. P.** 1974. Problem of protection of bats. *Conference material on bats, Leningrad, 1974*, 49-55.

213. **Topál, G. A.** 1962. Some experiences and results of bat banding in Hungary. Symposium Theriologicum. *Proc. int. Symp. Methods of Mammalogical Investigation, Brno, 1960*, 339–344.

214. **Topál. G. A.** 1966. Some observations on the nocturnal activity of bats in Hungary. *Vertebr. hung.*, **8**, 139–165.

215. **Topál, G. A.** 1976. New records of *Vespertilio murinus* (Linnaeus) and of *Nyctalus lasiopterus* (Schreber) in Hungary. *Vertebr. hung.*, **17**, 9–13.

216. **Tress, C., Henkel, F. & Haensel, J.** 1980. Kleine Mitteilungen. *Nyctalus*, n.s. **1**, (3), 263–267.

217. **Tupinier, D.** 1978. Gites artificiels pour chauve-souris. *Courr. Nature*, no. 56, 6–8.

218. **Tupinier, Y.** 1965. Chiroptères cavernicoles des Monts-cantabriques (Espagne). *Bull. mens. Soc. linn. Lyons*, **34**, 220–227.

219. **Tupinier, Y.** 1971. Les Chiroptères de la region Rhône-Alpes. *Proc. Congr. Suisse, 4th, Neuchatel, 1970*, 205–212.

220. **Tupinier, Y.** 1971. *Plecotus auritus* (L.) et *P. austriacus* (Fischer) dans la region lyonnais (Chiroptera Vespertilionidae). *Comptes r. Congr. nat. Soc. sav. Paris Sect. Sci.*, **96**, 227–234.

221. **Tupinier, Y.** 1975. *Chiroptères d'espagne systematique - Biogeographie.* Doctoral thesis, Universite Claud Bernard (Lyon).

222. **Tupinier, Y.** 1977. Description d'une chauve-souris nouvelle: *Myotis nathalinae* nov. sp. (Chiroptera - Vespertilionidae). *Mammalia*, **41**, (3), 327–340.

223. **Tupinier, Y. & Aellen, V.** 1978. Presence de *Myotis brandti* (Eversmann, 1845) (Chiroptera) en France et en Suisse. *Revue Suisse zool.*, **85**, (2), 449–456.

224. **Tupinier, Y. & Pontille, H.** 1971. Chiroptéres de la vallée de l'Azerques et des monts de Beaujolais (Department du Rhone). *Bull. mens. Soc. linn. Lyons*, **40**, (1), 24–28.

225. **Tuttle, M. D. & Stevenson, D. E.** 1978. Variation in the cave environment and its biological implications, *Proc. Symp. Nat. Cave Manage., 1977*, 108–121.

226. **Uspenskij, G. A., ed.** 1979. *Animal world of Moldavia: mammals.* Shtiintsa, Kishinev.

227. **Vernier, E.** 1978. I chirotteri del vicentino. *Stalactite, Bollettino Gruppo Grotte Schio*, 1–4.

141

228. **Vierhaus, M.** 1979. Nordfledermäuse *Eptesicus nilssoni* (Keyserling und Blasius 1839) überwintern im sudwestfalischen Bergland. *Z. Saugetierk.*, **44,** 179–181.

229. **Vierhaus, H.** 1982. Über einen weiteren Nachweis der Rauhhautfledermaus *(Pipistrellus nathusii)* aus Schleswig-Holstein und neue Unterscheidungsmerkmale zwischen Rauhhautund Zwergfledermaus. *Nyctalus,* n.s. **1,** (4/5), 307–312.

230. **Vierhaus, H. & Feldmann, R.** 1980. Ein sauerlandischer Nachweis der Nordfledermaus *(Eptesicus nilssoni)* aus dem winter 1972/73. *Nat. Heim.,* **40,** 97–99.

231. **Voûte, A. M.** 1980. The pond bat *(Myotis dasycneme,* Boie 1825) an endangered bat species in northwestern Europe. *Proc. int. Bat Res. Conf., 5th, 1980,* 185–192.

232. **Voûte, A. M.** 1981. Inheemse Vleermuissoorten: activiteiten ter bescherming. *Panda,* **17,** 146–148.

233. **Voûte, A. M., Sluiter, J W. & Heerdt, P. F. van.** 1980. De vleermuizenstand in einige zuidlimburgse groeven sedert 1942. *Lutra,* **22,** 18–34.

234. **Walravens, E.** 1980. Proportion de Chiroptères dans des pelores de rejection de *Strix aluco* en forêt de Soignes. *Bull. Inst. r. Sci. nat. Belg.,* no. 6, 62.

235. **Weber, B.** 1972. Weitere Fledermausnachweise für den Bezirk Magdeburg. *Nyctalus,* no. 4, 16–21.

236. **Wilhelm, M.** 1973. Zur Fledermausfauna der Slowakei. *Nyctalus,* no. 5, 26–28.

237. **Woloszyn, B. W.** 1981. Nietoperze i cywilizacja. *Rocz. Muz. Czest.,* **11,** 97–108.

238. **Wijngaarden, A. van, Loon, V. van & Trommel, M. D. M.** 1971. De verspreiding van de Nederlandse zoogdieren. *Lutra,* **13,** 1–41.

239. **Zingg, P. E.** 1982. *Die Fledermäuse (Mammalia, Chiroptera) der Kantone, Bern, Freiburg, Jura und Solothurn.* Dissertation, University of Bern.

240. **Zwinenberg, A. J.** 1979. Vleermuizen. **AO,** no. 1785, 1–20.

241. **Balcells, E.** 1955. Quiropteros del territorio Espanol: 3ª nota. *Speleon,* **6,** 73–86.

242. **Balcells, E.** 1956. Estudio biologico y biometrico de *Myotis nattereri* (Chir. Vespertilionidae). *Publnes Inst. Biol. apl., Barcelona,* **23,** 37–81.

243. **Carol, A., Samarra, F. J. & Balcells, R.** 1983. Revision faunistica de los Murciélagos del pirineo oriental y catalunya. *Monogr. Inst. Estud. Piren.,* **112,** 1–106.

244. **Kraus, M. & Gauckler, A.** 1980. Zur Abnähme der Kleinen Hufeisennase *(Rhinolophus hipposideros)* in den Winterquartieren der Frankenalb (Nordbayern) zwischen 1958 und 1980. *Myotis,* **17,** 3–12.